Closing the Gap

Closing the Gap

The Fourth Industrial Revolution in Africa

Tshilidzi Marwala

MACMILLAN

First published in 2020
by Pan Macmillan South Africa
Private Bag X19
Northlands
Johannesburg
2116

www.panmacmillan.co.za

ISBN 978-1-77010-786-1
e-ISBN 978-1-77010-787-8

Editing by Sean Fraser and Wesley Thompson
Proofreading by Sally Hines
Design and typesetting by Triple M Design
Cover design by publicide
Author photograph by Jan Potgieter

Contents

Introduction

In 2016, I delivered the 65th Bernard Price Memorial Lecture, and the topic I chose was 'Fourth Industrial Revolution and Society'. The reason I chose this as the focus was that I was concerned about how Africa was falling behind as a result of the exponential growth in technology. I felt then, as I do now, that we cannot afford for this trend to persist.

When President Cyril Ramaphosa mentioned the fourth industrial revolution (4IR) in his State of the Nation Address (SONA) in 2019, many dismissed it as mere rhetoric – no more than padding for the New Dawn. Just a few years ago, the 4IR seemed little more than a few buzzwords strung together. But, as we have witnessed through the rapid changes in our society, it's clear that the 4IR is no longer an abstract concept – it is, in fact, our lived reality. And although it still sounds somewhat fantastic – the confluence of the physical, digital and biological spheres through artificial intelligence (AI), automation, biotechnology, nanotechnology and communications technologies – the 4IR has already impacted the economy, citizens, society and the state. Technologies and processes are evolving at an exponential pace and are becoming increasingly inter-related. Substantial disruptions will affect all industries and entire systems of production, management and governance, and will undoubtedly transform all aspects of 21st-century life and society. Unlike the negative connotation of 'disruption' in the traditional sense, here it entails the innovations that make products and services more accessible and affordable for a larger population.

The 4IR, of course, succeeded other industrial revolutions; the first gave us steam engines, the second electricity and the mass production of goods, and the third digital technologies. With the 4IR, there has been a significant paradigm shift in every aspect of our lives – and this is just the beginning.

Africa is diverse. Yet, across the continent, we face the same challenges. Unemployment remains a scourge, economic growth remains lacklustre, and poverty seeps through every nation. This, however, need not be our reality. In September 2019, it emerged from the World Economic Forum (WEF) on Africa, held in Cape Town, that the 4IR has the potential to turbocharge socio-economic development across the continent. I would argue that to be skilled in areas such as AI, machine learning and blockchain – the technology used in digital currency – is key to tackling the issues of exclusion and social dispersion, while it is also crucial in fostering national unity. Of course, I would be remiss if I did not acknowledge that it also has the potential to exacerbate poverty and inequality. However, in heeding its call, we may also be able to subvert those issues. We cannot cower from the 4IR in fear that it may widen our disparities. Instead, we need to be prepared for the shift so that we can deploy technology in a way that is beneficial to all facets of society.

Across the globe, progress has been defined by the ability of humans to adapt to change. Here in Africa, we have largely missed that call. We have seen what a knock the continent has taken in being late to the previous three industrial revolutions – where gaps in infrastructure still exist today; where we have not yet been able to unleash our potential. We are now, however, uniquely poised to lead the charge.

For several years, I have been looking at how technologies will change the world of work in the 4IR. At the University of Cambridge, from 1997 to 2000, I used AI to analyse the performance of machines. In 2018, when I took over the reins as vice-chancellor and principal of the University of Johannesburg (UJ), I embarked on a journey to position the university at the forefront of the 4IR. The goal of UJ is to foster a platform for innovative and ground-breaking research and to produce graduates who are agile and curious and able to be active participants in a technology-driven and digital environment. Since then, I have been appointed as deputy chair of the Presidential Commission on the Fourth Industrial Revolution (PC4IR), to assist the government in taking advantage of the opportunities presented by the 4IR. To be effective, this, of course, requires collaboration between the government, the private sector, the unions and civil society. It was thus vital for me to piece together how the 4IR is impacting our lives.

This book begins as a journey tracing how the 4IR unfolded, how it impacts the world around us, and where South Africa and the continent fit in. This volume is divided into four parts, with a special lens cast on the impact of the 4IR on industry, data, business and society in Africa. Each chapter maps out a 4IR narrative exploring and probing the disruptions that have taken place, some rapidly and others in a more iterative and measured way. The areas covered in relation to the impact of the 4IR on African industries include the safety of structures (such as buildings), aerospace, mining and electricity. The topics covered in relation to the 4IR and data include data privacy, digital heritage, cybersecurity, social networks and 5G technology. The impact of the 4IR on African business includes topics such as economics, banking, taxation, market efficiency, trade and leadership. The matters under discussion in terms of the 4IR's impact on African society include languages, ethics, democracy, the movie industry, work, rationality, sports and memory.

It is envisaged that on this journey traversing different sectors, and by both describing and analysing the shifts the 4IR has already made, valuable lessons can be gleaned for how we, as a society, can embrace these changes and adapt with the required flexibility and agility. For it is only through understanding these vital footprints that we can take our place within this revolution and not be mere spectators.

CHAPTER 1

Understanding the fourth industrial revolution

The start of 2020 was awash with opinion pieces welcoming the 4IR. While many have argued that the 4IR is an abstract concept, it is becoming more apparent that it is not an era that looms in the future but rather our current reality, pervading our daily lives. As David Mills, CEO of electronics giant Ricoh Europe, put it in an article in *Forbes* magazine: 'I'm embracing the fourth industrial revolution with open arms.'[1] So, although the importance of the 4IR might not have evaded policymakers, what is still lacking is a real understanding of what it is. When the Presidential Commission on the Fourth Industrial Revolution (PC4IR) was established by President Ramaphosa, elsewhere arguments over its relevance and what the 4IR entailed for South Africa raged on. The commission was established midway through 2019, with President Ramaphosa as the chairperson, me as his deputy, and 30 other members drawn from diverse backgrounds, ranging from government to industry and labour. The task we were faced with was an urgent one: assist the 'government in taking advantage of the opportunities presented by the [4IR]'. This is no easy feat when you consider that there is a fundamental misunderstanding of the 4IR.

This could not be more evident than it was in parliament on 22 October 2019 when Deputy President David Mabuza was asked by a member of the Economic Freedom Fighters, Dr Mbuyiseni Ndlozi, to define the 4IR, simply because there is very little understanding of it in the country. The deputy president's answer was glaringly inadequate. Realising that the deputy president was struggling to answer, the Chief Whip of the Democratic Alliance at the time, John Steenhuisen, quipped: 'The deputy president needs to phone a friend.' This was not the first time that the deputy president had stumbled over a question about the 4IR, but it was perhaps in this moment that the gap in knowledge across not only South Africa but also among our government officials became apparent. Two days later, an article written by the deputy president was published in *Business Day* newspaper, in which he stated that the disparity in access to technology in South Africa is not something to joke about. This, of course, is not the only hurdle. According to a 2017 report by Dr Saahier Parker of the Human Sciences Research Council (HSRC), science and technology literacy in South Africa is very low. This also applies to our politicians. It seems futile then for us to criticise the deputy president for his seemingly incomplete understanding of the 4IR when there is a fundamental need to put mechanisms in place to increase scientific literacy in our society.

As Pieter Geldenhuys remarked on Moneyweb, many people holding responsible positions in South Africa do not understand the 4IR. More concerning is that all these 4IR technologies are fundamentally changing our generation. So, had the deputy president phoned a knowledgeable friend, how would they have defined the 4IR? For us to understand the 4IR, the first task is to be familiar with the history of industrial revolutions. The 4IR is the fourth because there were first, second and third industrial revolutions. It is essential to know where South Africa has fallen behind in each industrial revolution and why it is so necessary that we do not suffer the same fate a fourth time.

The first industrial revolution

The first industrial revolution began in England in the 18th century. Statistically, of course, it should have happened in China or India because these two nations, just as they do today, had the largest populations. There is no consensus as to why England led the way. The first industrial revolution was inspired by the scientific revolution, part of the Renaissance, which saw significant developments in mathematics, physics, astronomy, biology and chemistry. The scientific revolution had, in turn, been catalysed by the Reformation, a movement within Western Christianity 'that posed a religious and political challenge to the Roman Catholic Church and in particular to papal authority'.[2] The Reformation was started in Germany in the 16th century by Martin Luther, and it introduced a culture of individualism and personal responsibility. As the Reformation increased in intensity, particularly in England, it became a significant driver of the scientific revolution, which then brought about the first industrial revolution.

During this time, many agrarian and rural societies in Britain became industrialised and urbanised. Alongside the rapid scientific, technological and commercial innovations were a rising population, improved transportation and expanding domestic and international markets.[3] Employment opportunities could mainly be found in the mining, engineering and manufacturing sectors. It was the first industrial revolution that gave rise to steam engines and the mechanisation of the production of goods. Until then, manufacturing was done by hand by trained craftsmen, but the first industrial revolution allowed for commodities to be manufactured in bulk in factories, and steam trains gave rise to railroads that carried massive amounts of goods far more quickly and efficiently than could be done on horseback, for example. The first steam train arrived in South Africa in 1860, some 60 years after the world's first steam train was built. But, even at the time, no nation could afford to be 60 years behind.

The first industrial revolution resulted in an increased volume and variety of manufactured goods and an improved standard of living – but only for some. It also, of course, resulted in grim living conditions and unemployment for the poor.[4] In the textile industry, for instance, machines began to replace human labour. This development was met with fierce opposition from the Luddites, groups of English workers who organised into a form of a labour union and who set out to destroy any machinery they believed would threaten their jobs. In retaliation, many were arrested and executed. Yet, despite the opposition of the Luddites, the first industrial revolution marched on, and the Luddites faded into obscurity.

At this time, the South African context was vastly different. The Dutch held colonial control of South Africa, only for the British Empire to inherit the Cape during the Napoleonic Wars. From the first industrial revolution, leading into the second, the discovery of diamonds in 1867, as well as gold in 1886 in South Africa, transformed the economic and political structures of the country. The discovery that South Africa's diamond and gold resources exceeded those in any other part of the world naturally attracted foreign capital and large-scale immigration. And, of course, for this to be profitable, an enormous amount of inexpensive labour was required, which was made possible as the British colonised more African states in southern Africa in the 1870s and 1880s, effectively taking control of much of the land and imposing cash-taxation laws.

The second industrial revolution

The second industrial revolution took place mainly in the United States, and it was based on the ideas of electromagnetism formulated by Michael Faraday and James Clerk Maxwell of Britain in the 19th century. Their theory was relatively simple, but proved to be a game-changer. When a person places an object that can conduct electricity, such as a piece of copper wire, next to a magnet, and the person moves this object, then electricity is generated. The reverse is also true: when we pass electricity through a conductor located next to a magnet, the conductor moves, and this is the basis of an electric motor. The concept of moving a conductor to generate electrical power, or of creating mechanical energy by passing a current through a conductor, unites magnetic and electrical forces and is called electromagnetism.

So it was that electromagnetism gave us electricity and the electric motor. This, in turn, spurred the second industrial revolution, which saw the introduction of electricity and mass production and changed the scale and speed of manufacturing significantly. Here, the assembly line of production was introduced, ushering in mass production. Perhaps the best-known example of the use of an assembly line is that which was done by the Ford Motor Company. While Henry Ford did not invent the automobile, he brought about the first mass production of cars. The fears we see now around the 4IR were not uncommon then. For instance, even Charlie Chaplin's assembly line comedy *Modern Times* depicts his Little Tramp character struggling to survive as a factory worker in the modern, industrialised world.

At the same time, South Africa's industrialisation had become increasingly dependent on what is referred to as the Minerals-Energy Complex. State-owned corporations in the 1920s – electricity, steel and transport companies – worked in tandem 'with private mining conglomerates to build an economy based around the extraction of minerals using cheap forms of energy'.[5] This was made possible by racial segregation in South Africa, which ensured a never-ending supply of cheap black labour. As Khadija Sharife and Patrick Bond put it, 'The Truth and Reconciliation Commission [TRC] determined after the 1996 hearings that the South African mining industry's direct involvement with the state in the formulation of oppressive policies or practices that resulted in low labour costs (or otherwise boosted profits) can be described as first-order involvement [in apartheid].'[6]

This, of course, has had a lasting legacy, one that we have not been able to move beyond. Eskom, South Africa's power utility, still produces coal-generated electricity, which has proved to be an inadequate energy strategy for a country that has seen rolling blackouts in the last decade. In South Africa, there is a distinct lack of the imaginative diversification that is evident in other countries, such as India, where Cochin International Airport in the south is the world's first solar-powered airport.

The third industrial revolution

The third industrial revolution came about following the invention of semiconductor devices, such as the transistor, in the 1940s and 1950s. Semiconductors are materials that conduct electricity under specific conditions, and they were central to the invention of the transistor in 1947, which ushered in the electronic age. Transistors power our phones, computers and televisions. In technology, analogue thus moved to digital. For example, old television sets that were tuned by antennae have now been replaced by tablets and other hand-held devices, such as the iPad, which are connected to the internet and allow you to stream movies. How does this work? Because transistors conduct electricity under certain circumstances, they can be used as efficient switches to switch on (1) or off (0). This switching on and off makes them suitable digital machines. In other words, a small electric current flowing in one portion of a transistor can cause a much greater flow of current in another section. This is essentially how all computer chips work.[7] An example of the use of a transistor is in hearing aids, in which small microphones detect sounds, which are turned into fluctuating electrical currents that are in turn inputted into a transistor, which amplifies them through a speaker, resulting in the wearer hearing a much louder version of the small sounds.[8]

The idea of using ones and zeros to communicate information did not originate with digital computers, but was used earlier in devices such as the telegraph. Telegraphs were in use until the early 1990s and used a dedicated channel to transmit messages. I last received a telegram in 1991 when I was studying for an undergraduate degree in Mechanical Engineering at Case Western Reserve University in the US, which stated that 'Tshianeo is dead STOP'. As the telegram used a dedicated channel, it was costly, and so it had to be as brief as possible. The message I received in 1991 was informing me that my grandmother, Vho-Tshianeo, had died. My family had to drop the prefix 'Vho' from her name to save on costs. The discovery of efficient switching machines in semiconductor devices automated the sending of telegrams and thus ushered in the electronic age.

Semiconductors are used in the creation of transistors, the basic unit of the digital conductor. To comprehend the digital industrial revolution (another way of referring to the third industrial revolution), we need to understand the concept of digital and what it means. When we write words, we use 26 letters of the alphabet. In this regard, we can write the entire isiZulu or English language using 26 letters. Furthermore, we can write all the numbers using only ten characters, 0 to 9. The idea of writing the entire language with 26 characters should not be taken lightly, because languages such as Chinese require several thousand characters. The digital industrial revolution allowed us to express all information, whether pictures or language or numbers, with just two characters: 1 and 0. This is done using what is known as 'binary code', which is essentially a coding system in which each letter or number is assigned a code made up of a string of ones and zeros. Computers contain billions of transistors, each of which can be switched on (1) or off (0), and which can, therefore, be used to store information as binary code.

However, in spite of the invention of the revolutionary digital technologies made possible by the use of semiconductors and transistors, not everyone across the globe benefited in the same way. Some countries, such as South Africa, lacked widespread and equal access to these digital technologies. Rampant inequality in the country has resulted not only in disparities in access to technology, but also in a lack of innovation. To this day, 70 years after the invention of the transistor, in South Africa, we have neither a home-grown computer nor a semiconductor industry.

The fourth industrial revolution

The 4IR is a confluence of cyber, physical and biological technologies. But the 4IR is, of course, not without its challenges. Some are afraid that the 4IR will lead to talent shortages simply because many do not have the necessary skills to adapt to the disruptions in industries, mass unemployment and growing inequality. A 2018 Accenture study concluded that approximately six million jobs in South Africa would be at risk of being lost to automation by 2025. This study showed that both blue- and white-collar jobs are in jeopardy, including those of clerks, cashiers, bank tellers, construction workers, mining and maintenance staff.

Some of the elements that drive the 4IR include AI, robotics, the internet of things (IoT) – a network of objects connected by the internet – and gene editing. AI is a technology that brings intelligence to machines. Alan Turing, essentially the founder of AI, developed the Turing Test, which suggests that a machine is considered 'intelligent' if the human who interacts with it cannot immediately tell whether what they are up against is machine or human. And yet, although we have not yet built a machine that can consistently pass the Turing Test, AI is still automating a large number of processes around us. As a result of AI, the best chess player in the world is a machine called Deep Blue, and not a human being. Because of AI, the best player of the Chinese game called Go is a machine named AlphaGo, created by Google's DeepMind, and not a human being. In medical services, AI is reading medical images better than human radiologists. Because of AI, the world of work is shrinking while our productivity is increasing. So, while the first, second and third industrial revolutions gave us many high-paying industrial jobs, the fourth industrial revolution is providing us with a mass of unemployed and unemployable people.

But, if AI is the most dominant technology in the 4IR, what is it actually? AI is a technique that essentially makes machines intelligent. While computers traditionally relied on people to tell them what to do and how to react, AI means that machines can learn and make their own decisions.

There are three types of AI: machine learning, soft computing and computational intelligence. Machine learning is an approach to building intelligent machines that use statistics and data to automatically improve the way they function, in the same way that a human would use sensory and other data in order to learn. An example of machine learning is neural networks, which are collections of algorithms that mimic how the human brain works, particularly in terms of recognising patterns. Just as humans learn patterns, so too can AI be trained to recognise patterns, and this can be used to improve how they perform certain tasks. For example, if you were to touch a hot metal object, your immediate reaction would be to pull your hand away quickly. The lesson is usually learned due to the experience of pain, which influences your future behaviour. This sequence of events – as well as the memory of how it felt to burn your hand – is stored in your brain, reminding you not to repeat this action. This knowledge means that the next time you encounter a metal object you know will be hot, you are unlikely to touch it. This is how human intelligence works. In much the same way, AI is based on machines learning patterns and mimicking human intelligence and, in some instances, even surpassing it. The basic idea behind AI is to see whether we can give computers some of the decision-making abilities that we as humans have. The neural network is illustrated in Figure 1.1.

Neural networks that are large, with multiple layers, such as the one illustrated in Figure 1.1, are called deep learning. Neural networks were first proposed in the 1940s by Warren McCulloch and Walter Pitts, and they were perfected in the 1980s by researchers such as Geoffrey Hinton and John Hopfield. Today, machine learning is the most successful AI method. We now have an extensive capability to gather vast amounts of data, and we have powerful computers that can analyse this data and teach themselves how to learn from it in order to improve their efficiency and effectiveness.

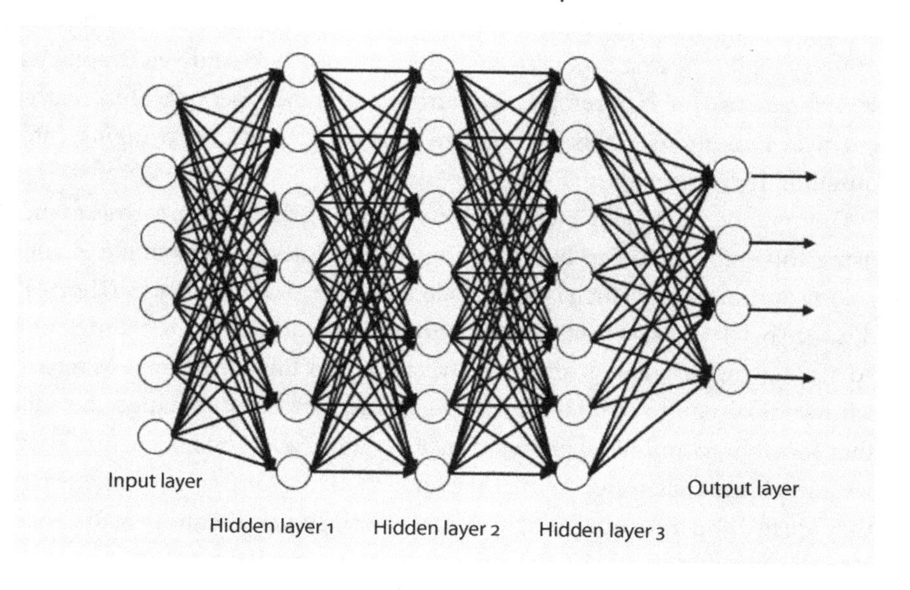

Figure 1.1 *A neural network*

Figure 1.1 shows how a neural network is comprised of layers of nodes, or 'artificial neurons' that, like synapses in the human brain, connect to each other in order to form a network by transmitting information to each other. The input layer receives the inputs, while the output layer delivers the final results. The hidden layers allow for the overall task to be broken down, with each layer being used to perform a certain part. For example, in a facial-recognition system that uses neural networks, the functions in one hidden layer used to identify eyes might be used in conjunction with other layers to identify other facial features in images.[9]

Soft computing is a technique of bringing intelligence to machines by using a limited amount of data. An example of soft computing is fuzzy logic. Fuzzy logic is based on vague, imprecise notions that may be more or less true – in other words, it is a logic based on partial degrees of truth. For example, traditional logic would tell you that if it is cloudy, then it will rain, and if it's a bird, it can fly. Fuzzy logic is more complex. It takes into account that some birds, such as chickens or ostriches, cannot fly. Fuzzy logic suggests that if there are clouds, then it will *probably* rain, but that something else may well happen; if it's a bird, then it can *probably* fly, but it could also be a flightless or even an injured bird. With fuzzy logic, one can extract expertise from a human being and transfer that to a machine. The notion of fuzzy logic was first espoused by Lotfi Zadeh in 1965 and yet, despite its widespread use today, it remains a controversial subject because of its mathematical limitations when it comes to providing solutions, bringing into question its consistency and validity.

Computational intelligence involves observing how nature works and using this information to build intelligent machines. One example is ant-colony optimisation, which is used in electronic maps to identify the shortest distance between two points. Ant-colony optimisation has its roots in South Africa, having originated with the Afrikaans poet Eugène Marais. Marais – who would go on to write the book *Die Siel van die Mier* – studied the ants that form large anthills, illustrated in Figure 1.2, and concluded that they were intelligent. He observed that they operated by emitting pheromones. If they are moving from one point to another, for instance, the shortest distance between the two locations will be the one with the most potent pheromones. Marais, however, published his theories in the *Huisgenoot* magazine in the 1920s,[10] and the work was subsequently stolen by Belgian Nobel Prize winner Maurice Maeterlinck.[11]

One of the confusing aspects of industrial revolutions is discerning the difference between the third and the fourth industrial revolutions. While the digital industrial revolution is about digitising our infrastructure, the

Figure 1.2 *An anthill*

4IR is about reconstructing our infrastructure to be intelligent. While the digital industrial revolution gave us computers, the 4IR gives us computational forms with which we can interact using speech. And whereas the digital industrial revolution automated production, the 4IR provides intelligent automation whereby machines can repair themselves. While the digital industrial revolution is based on digital computing, the 4IR blends digital computing with quantum computing, based on servers that use enormous amounts of information and less energy. Thus, the fourth and digital industrial revolutions are quite different.

In South Africa, while we are modernising our digital industrial revolution infrastructure such as television, we should thus not be left behind by the 4IR. The arrow of progress incorporates all the success that has been made earlier in the first, second and third industrial revolutions and so our focus must now be on the 4IR.

The 4IR globally

South Africa is arguably playing a challenging game of catch-up here. Many other nations have already begun planning for the 4IR. At the beginning of 2020, *Forbes* magazine stated that China is well on its way to becoming the first AI superpower. While over twenty nations already have AI strategies in place, China is by far the most ambitious. In 2017, the Chinese government unveiled their 'Next-Generation Artificial Intelligence Development Plan', which positions China securely as a leader in matters of the 4IR. China has already looked at the provision of legal frameworks, resources and goals, coupled with local freedom to adapt, for the 4IR, and expects to have implemented its strategy by 2025. Chinese President Xi Jinping said in 2018 that the leaders of the Communist Party of China need to 'ensure that our country marches in the front ranks when it comes to theoretical research in this important area of AI, and occupies the high ground in critical and AI core technologies'.[12] In large part, China has been able to make great strides in AI because of its political system and the government's push for technological advancement. For example, China has been using and refining a system to keep track of the behaviour of its citizens using AI. This is called the Social Credit System. For example, if citizens litter, they may lose points, but if they pay their bills on time, they may add to their score. The government then ensures that everyone is rewarded for good behaviour. Banks could offer people with high social scores lower interest rates, for instance, or reward them with faster promotion and better job offers. China is not alone in using this model – Singapore has had similar success. Elsewhere, China has collaborated with private-sector players by offering tax breaks and incentives, and it has worked with large conglomerates such as e-commerce giant Alibaba and tech companies like Baidu or Tencent. The government has also pushed to attract talent from outside China to ensure that it stays ahead of the curve.

The impact has been enormous. China has taken over from the US as the most significant capital market for AI start-ups and now publishes the most research papers in the field.[13] The landscape in the US is somewhat different. In 2016, the White House released a document titled 'Artificial Intelligence, Automation, and the Economy' in an effort to position itself as a leader in the fourth industrial age. In 2018, US President Donald Trump signed an executive order encouraging government agencies to do more with AI, suggesting that this is paramount in protecting the economic and national security of the country. The US, however, has been criticised for not planning for or investing in attracting talent in the AI space. China's model, of course, would not be possible in a capitalist-centred country such as the US, which has stringent privacy legislation in place. This AI race has already begun to play itself out in trade tensions between the world's two largest economies. While the US has put some Chinese AI companies such as Huawei on sanctions lists, the export of American AI software to China has also been restricted. While the US and China are in the process of signing a trade deal that serves as something of a ceasefire, the impact over two years has hit markets and stunted global growth.

But it is not just these two economies at play in the AI space. In 2017, Germany released 'Germany Industry 4.0', setting out a plan to use AI to build an industrial base grounded in the IoT in order to automate all aspects of the economy efficiently. In 2018, India adopted the 'National Strategy on Artificial Intelligence' to develop research ecosystems, address the skills deficit and promote the adoption and applications of AI. In 2017, the United Arab Emirates (UAE) appointed a minister of state for AI and became the first country in the Middle East to launch an AI strategy, the government announcing in 2019 that it would pilot the first university to have a singular focus on AI. The UAE's AI strategy will concentrate on developing a workforce skilled in the rapidly changing technologies that are transforming economies globally.

While it is difficult to group South Africa with the pool of countries already well on their way in the development of AI, our country is also lagging behind those on the continent. In 2018, the Kenyan government put together a Blockchain & Artificial Intelligence Taskforce to come up with 'recommendations about how to harness these emerging technologies over a five-year period'.[14] In a report in July 2019, it recommended implementing blockchain technology in areas such as service delivery, agriculture and health. It also recommended curbing the country's national debt through digital-asset frameworks – where content is stored digitally – and establishing a regulatory framework.

The 4IR in South Africa

So, where do we stand? Given what is happening globally, South Africa has indeed begun to develop its own 4IR strategy. However, whereas the approach common to so many other countries is based primarily on AI, the South African 4IR plan includes the third or digital industrial revolution because that is the one revolution that the state has not yet succeeded in building. There is, though, already evidence that the 4IR is rapidly changing industries in South Africa. For instance, in the manufacturing sector, operators are able to adjust the behaviour of robots in order to increase their capacity to support human operation and safety standards. By the 2000s, the Ibhayi Brewery in Port Elizabeth was already the most automated in the southern hemisphere, but then, in 2018, South African Breweries invested R438 million to expand and further automate the brewery.

This move towards automation in South Africa has not gone unnoticed, and the work of the PC4IR has had to be swift. At the end of 2019, we drafted a preliminary report outlining eight recommendations. The report was presented to the cabinet for consideration – a matter of great urgency when we consider the strides already made by other countries. Some recommendations include: investing in human capital; establishing an AI Institute; establishing a platform for advanced manufacturing and new materials (these are new materials used in manufacturing that are engineered in order to ensure they are, for example, lighter, stronger or more adaptable); securing and availing data to enable innovation; incentivising future industries, platforms and applications of 4IR technologies; building 4IR infrastructure; reviewing and amending or creating policy and legislation; and establishing a 4IR Strategy Implementation Coordination Council in the Presidency.

We will delve deeper into each of these recommendations in Chapter 3.

While creating a strategy was an essential first step, South Africa needs to ensure that this is implemented urgently for us to compete with the likes of Kenya, let alone China or the US. This will not only require collaboration between all spheres of society, ranging from government to labour to the private sector, but it will also – in a show of political will – need to be kick-started by parliament. If the previous three industrial revolutions have taught us anything, it is that we cannot afford to miss the boat again if we are to emerge with lower unemployment rates and higher growth.

CHAPTER 2

Automation

In 2019, Lee Se-dol, the South Korean champion of Go, announced that he was retiring from professional play because of AI, declaring that AI was unbeatable. Go is a strategy board game played on a grid where a player needs to surround more territory with stones than their opponent. Invented in China 2 500 years ago, it is the oldest board game consistently played until the present day. For many years, Go was considered beyond the reach of even the most sophisticated computer programs. Back in 2016, AlphaGo, developed by Google-owned AI company DeepMind, proved a worthy adversary, defeating Se-dol four matches to one. The question then becomes, if AI is beating humans in games such as Go, in which other areas is it surpassing or even replacing us? Substantial disruptions as a result of AI will affect all industries and entire systems of production, management and governance, and it will undoubtedly transform all aspects of 21st-century life and society. The fears around the 4IR are somewhat different. It is no longer just blue-collar jobs at risk, but white-collar ones too.

Automation and AI

The threat to jobs in the 4IR is mainly as a result of automation, the process of using machines to perform tasks usually done by humans. But automation is not a new concept. It is, in fact, ancient, and can be traced back to 270BC when, in Egypt, water clocks were used to tell the time, automating what until then had exclusively been done by humans using the passage of the sun and the moon. Because the sun is not always visible, meaning that sometimes people couldn't tell the time, the water clock became essential. In the first industrial revolution, the steam engine was used to automate manufacturing. In the second industrial revolution, an electric motor was used to move the conveyor belt and produce goods, such as cars, at an astonishing rate. The result was automated production on a larger scale than what could have been achieved in the first industrial revolution. The third industrial revolution combined the transistor and the electric motor of the second industrial revolution to automate industries electronically. As noted in Chapter 1, the transistor is the basis of the electronic age, and, when we use it in integrated circuits, we can build complicated robots and improve productivity through automation. In the 4IR, these machines are becoming intelligent through the use of AI. We can consider a machine intelligent if it is able to analyse information and extract insights beyond the obvious. Machines of the 4IR are, therefore, much more useful than machines of the third industrial revolution. They can self-repair and make decisions without the intervention of a human being. In fact, machines are becoming superhuman and are starting to change entire industries.

Through using AI, we are able to make automation intelligent. Through AI, companies are able to predict equipment malfunctions, supervise workers, and increase productivity without the aid of humans. In Johannesburg, an AI-based system that can foretell the failure of electrical transformers has been constructed to improve electricity supply. Transformers are vital and expensive machines in our electricity grid – they, for instance, allow us to draw electricity from the network to power household devices, such as televisions and fridges. This reduces the equipment required to power devices as well as increases efficiency and productivity.

South Africa has also produced an AI system that can help people who have lost their voice regain the ability to speak. The *MIT Technology Review* covered the research, for which a patent was registered via the Patent Cooperation Treaty. We have also used AI to classify the various strains of leukaemia, to predict epileptic episodes, and to analyse images of patients' lungs in order to diagnose pulmonary embolisms. The increased speed and accuracy of cancer diagnostics through analytics, which can autonomously characterise tumours and prescribe therapies, have not replaced doctors, but have accelerated their efforts and given them the time to attend to more patients. We have even used AI to monitor the state of repair or structural integrity of dams and bridges using low-cost cameras and sensors.

Young South African start-ups, such as Aerobotics, have since cemented their role in the 4IR. Aerobotics develops drone technology used in the agricultural, logistical and mining industries. This allows farmers, for example, to scan their farms and provide analytics in order to manage them more efficiently, thereby reducing costs and increasing yields. Its AI systems thus assist farmers in optimising the utilisation of their land and in decreasing monthly water, fertiliser and diesel costs.

What is to be done about automation?

Given all these advances, what does the future hold for humanity? To begin with, the world of work will shrink as intelligent machines take on more jobs. This, in turn, will increase inequality. When inequality rises, the aggregate demand decreases, thereby depressing the markets. Suggested remedies include the introduction of a universal basic income (UBI), but the reduction in the workforce will obviously result in a smaller tax base, thereby making the UBI unaffordable. In the first, second and third industrial revolutions, human beings invented trade unions to fight against exploitation in the workplace. The 4IR will usher in the problem of human irrelevance, which will have a significant impact on our identity. The 4IR's dependence on data is even altering the notion of democracy. Companies that collect vast amounts of data are using this to influence national elections and, thereby, are usurping power for the benefit of multinational companies.

Given all the challenges that automation brings, what is to be done? Firstly, we need to craft legislation that will regulate the technologies of the 4IR. This should regulate ownership of data so that people can be protected, and specify how platform companies doing business in South Africa, such as Uber, Facebook and Twitter, should be regulated and taxed.

Secondly, we need to create a system that will incentivise the collection and ownership of data. Google Maps does not pronounce indigenous street names correctly because the data used to program these systems is gathered in North America, Europe and Asia. If we – as South Africans, as Africans – do not collect these data, then we shall be locked out of the 4IR.

Thirdly, we need to educate our people so that they understand all the developments happening around them. The way education was structured in the first, second and third industrial revolutions is not geared towards the 4IR, in which entirely different skills are required. Instead of requiring students to simply memorise facts, in the 4IR, we need them to develop critical-thinking skills. Instead of education that is too specialised, the 4IR requires that we include systems-thinking in our learning. Because the jobs that will remain relevant in this new era will require a human touch, understanding people and society are essential. For us to achieve all of this, we need to offer a multidisciplinary education system. Those who are studying technical subjects must understand human and social sciences, while those who are studying human and social sciences must understand technology. We also need to introduce computational and mathematical thinking in our curricula.

Job losses

While many dread that millions of jobs will become obsolete as a result of automation and AI, these technologies provide an opportunity to transform vocations and increase productivity. This is particularly significant when viewed in the context of South Africa's economy growing at less than 1%, with the unemployment rate at 29% and rising. Given the above, what will happen to jobs in the era of increasing automation? Will we still have the jobs we currently have in the age of the 4IR? Here, it is essential to reconsider the work of Hans Moravec, who suggests that the level of difficulty of automating a job depends on how long humans have been executing that task on an evolutionary timescale. For instance, humans have been climbing trees for tens of thousands of years and, consequently, it is incredibly challenging to automate climbing trees. However, human beings have been writing for only 6 000 years, so it is easier to automate writing than to automate climbing trees. In fact, because it is now so simple a task for a computer to read, as well as to check spelling and grammar, companies no longer employ people to perform these tasks, since software such as Microsoft Word or Grammarly is now capable of executing this. What Moravec proposed is that, in order to understand the consequences of intelligent machines taking over jobs, we need to understand that white-collar workers are more vulnerable than blue-collar workers. Therefore, it is not just unskilled workers who must worry about being replaced by machines, but also workers in skilled professions such as medicine or finance. Hence, automation is changing both blue-collar and white-collar jobs and replacing them with silver-collar jobs (those done by machines or human and machine).

Economist Kenneth Rogoff clarifies in an article for the WEF that 'since the dawn of the industrial age, a recurrent fear has been that technological change will spawn mass unemployment. Neoclassical economists predicted

that this would not happen because people would find other jobs
sibly after a long period of painful adjustment. By and large, that pr
has proven to be correct.'[1] For example, in the 18th century, fears about the impact of machinery on jobs intensified with the growth of mass unemployment. Nevertheless, in the second half of the 19th century, it became increasingly clear that technological advancement was indeed benefiting all segments of society, including the working class.

New jobs

Reports have suggested that while the 4IR will result in massive job losses, even making some careers obsolete, it will also pave the way for new silver-collar jobs, especially in the fields of science, technology, engineering, arts and maths (STEAM), among them data analysis, computer science, engineering and occupations in the social sciences. There are three broad factors that will define the 4IR. Firstly, some jobs will disappear altogether. Secondly, some jobs will change. An example of this is in the medical profession, where doctors will increasingly be required to be competent in technology. Thirdly, new jobs will emerge. For example, banks are now hiring Chief Artificial Intelligence Officers – a role that did not exist a few years ago.

The 2018 WEF report noted that the major disruptors in the workforce are fast internet, AI, big data – large volumes of data that can be analysed in order to provide insights – and Cloud technology. The characteristics of the workforce are indeed changing and shifting, with humans projected to only perform 58% of all task hours compared to machines executing 42% of these hours by 2022 (down from 71% and 29% respectively in 2018). It is predicted that the market will demand professionals who are able to blend science and technology with human and social sciences, readily achieved through multidisciplinary education. In nurturing these skills, universities are introducing innovative degrees, such as a Bachelor's degree in Politics, Economics and Technology, to fill the gap in the labour market.

The strategy of staying ahead of the changing needs of the labour market is to retrain, upskill and transform the workforce. Career paths are fundamentally changing. The move is, for example, towards hybrid jobs premised on combining skill sets, such as marketing and statistical analysis, or design and programming.

Dr Martyn Davies, MD of Deloitte's Emerging Markets & Africa, raises pertinent questions in an Investec Focus podcast: 'Are your jobs being taken? Maybe they have been displaced somewhere else. Jobs are not destroyed; they are just displaced. The challenge for us in our part of the world is – how do we capture the displacement?'[2] South Africans should thus position themselves strategically to lead the 4IR in Africa rather than playing catch-up. Here, universities play a vital part in crafting skills for the 4IR. The workforce should be armed with transferable skills through a broad range of job opportunities and workers should be assisted in adjusting their approaches to solving business problems in dynamic industrial environments.

The role of education

Given the above, how do businesses succeed in an increasingly changing business environment? Here, universities play a fundamental role in developing skills of the future generations as they navigate new technological directions, which, as we see now, requires the reskilling and upskilling of our workforce to close the skills gap. Furthermore, universities should play a critical role in developing effective intellectual property (IP) strategies to be implemented through the Technology Transfer Offices (TTOs). In this way, TTOs will be able to capture knowledge that can be commercialised. The TTOs must also be able to register and commercialise patents to grow the economy. UJ has, for instance, already positioned itself at the forefront of this change.

As the career landscape evolves, universities need to align their curricula with developments in automation in order to equip students for further study or employment. Amid the rapid and fundamental changes that are occurring in industry, universities need to revisit how knowledge is packaged into modules and qualifications. Many universities have started working together with businesses and are now incorporating the skills they require into the curriculum. Equally, many employers are now partnering with universities to develop tailored learning programmes for employees in order to prepare them for emerging job opportunities in the automated world. This is the challenge universities face: how to adapt the curriculum best to meet the needs of business.

So, how do universities help companies adapt to this changing, automated work environment? The answer must surely include restructuring our degree programmes. To acquire a broader range of skills, students need the freedom to choose courses outside of a programme, and universities must respond to this. Today, students want to determine the pace at which they learn and also have remote access to content. There was, for instance, an urgent shift towards remote learning as universities adapted to national lockdowns necessitated by the COVID-19 outbreak. Many universities are already adopting a model that allows students to enrol in stackable degrees – which are comprised of courses or modules from diverse faculties and disciplines that are stacked together to form an academic degree – through multiple entry and exit points and by using technology such as data analytics to develop customised learning paths. It has thus become pivotal for companies to leverage and invest in pertinent innovation and disruptive technologies. Universities can support this development by offering flexible multi-disciplinary courses that respond to these real-world needs. Our education should, therefore, introduce computational and mathematical thinking so that the learning outcomes facilitate the understanding of the critical drivers of the 4IR.

The role of universities is to prepare graduates for the workforce, but some of these graduates, of course, proceed to do postgraduate work or further study. This challenges the old-fashioned ways we view curricula and teaching methods. As the education system adapts to this changing landscape, part of our role has to include teaching our students about the 4IR. As Bruce Lee, martial-arts expert, put it in his book *Tao of Jeet Kune Do*: 'You must be shapeless, formless, like water. When you pour water in a cup, it becomes the cup. When you pour water in a bottle, it becomes the bottle. When you pour water in a teapot, it becomes the teapot. Water can drip, and it can crash. Become like water, my friend.'[3] For our economy to grow, our education system must become more adaptable.

The role of business

Given all these developments, how does business respond to automation? A report by Tata Consultancy Services (TCS) calls for companies to adopt four key behaviours. First, to drive mass personalisation, which allows companies to offer personalised products and services at scale. Second, to create exponential value, which empowers businesses to extract more benefit from a single transaction using technologies such as AI and big-data analytics to estimate future customer needs and points of engagement to target a broader range of potential customers. Third, to leverage ecosystems to foster collaboration among multiple partners beyond supply-chain networks so that they are no longer forced into silos. And, finally, all these behaviours require that companies embrace risk in order to stay ahead of the competition and offer enhanced value to customers. The TCS report found that only around one in ten organisations surveyed have adopted all four behaviours.[4]

The shift in the workplace is built on enhanced efficiency, as human beings and intelligent machines learn to work together, thus enhancing supply chains and eventually improving economic growth. This is particularly pertinent in South Africa, as we attempt to grow our economy. In a country that has not managed to achieve growth rates of at least 5% – the benchmark required to make a dent in the increasing unemployment rate – it is essential to tap into the opportunities that the 4IR brings. South Africa is a globalised, open economy, so weaker global growth harms us – as seen during the global financial crisis. While global growth remains tepid, there is untold potential into which companies can tap, a potential that the WEF approximates will create up to US$3.7 trillion in value by 2025.

As Economics Nobel Prize winner Robert J Shiller put it, 'You cannot wait until a house burns down to buy fire insurance on it. We cannot wait until there are massive dislocations in our society to prepare for the Fourth Industrial Revolution.'

According to the International Data Corporation's *IDC FutureScape: Worldwide IT Industry 2020 Predictions*, much of the spending on AI systems will be driven by the retail and banking industries. Approximately half of the retail spending will be on automated customer-service agents and expert shopping advisers, as well as recommendation systems. The banking industry will invest in automated threat-intelligence and prevention systems as well as fraud analysis and investigation to deal with cyber threats.[5]

CHAPTER 3

South Africa's 4IR strategy

At the beginning of 2020, a Deloitte report titled *The Fourth Industrial Revolution: At the Intersection of Readiness and Responsibility* was released ahead of the 50th WEF in Davos. It found that while C-suite executives – in other words, top senior management – were on board with the concept of Industry 4.0 (another term for the 4IR), many were yet to draft an action plan to turn goals into reality. Deloitte had surveyed more than 2 000 C-suite executives across nineteen countries to understand how much progress they had made with 4IR technology, and it found that there was a distinct disconnect in translating 4IR priorities into business practices. While more than twenty countries have AI strategies in place, many still lag behind.

And here, in South Africa, we are playing a challenging game of catch-up – one that, of course, goes beyond just AI, as we try to close the gap in the previous three industrial revolutions, where we have lagged behind. According to Accenture, harnessing digital technologies can generate R5 trillion in value for South African industries over the next decade, particularly in agriculture, infrastructures, manufacturing and financial services. As President Ramaphosa put it at the inaugural Digital Economy Summit in July 2019, 'We were left behind by the first industrial revolution, the second and so forth, but the fourth one is not going to leave us behind – we are going to get ahead of that fourth industrial revolution.'

South Africa's readiness for the 4IR

So, where does South Africa currently stand? According to the benchmarking framework of the WEF, which assesses a country's readiness for the 4IR, South Africa is categorised as being in the nascent quadrant, indicating that it is just beginning to develop. The assessment argues that South Africa's manufacturing share of GDP has decreased since the early 1990s to approximately 12% today, as its services sector has expanded. Despite this, the country has the most robust Structure of Production within Africa. However, across the Drivers of Production component, South Africa's performance is mixed. On the one hand, the ability to innovate is one of South Africa's greatest strengths, as the country has a strong culture of innovation and a sophisticated financial sector that supports entrepreneurial activity. On the other hand, human capital remains the most pressing challenge in preparing for the future of production, simply because there is a shortage of engineers and scientists as well as digital skills.

As the WEF's prognosis reads, 'It will also be critical for South Africa to improve its Institutional Framework to effectively respond to change, offer a stable policy environment and direct innovation.'[1] This is the context in which we are creating a framework for the 4IR in South Africa.

The Presidential Commission on the Fourth Industrial Revolution (PC4IR)

In 2019, President Ramaphosa established the PC4IR, which he chairs. The PC4IR was tasked with the responsibility of putting together a plan that will ensure that South Africa takes advantage of the opportunities presented by the 4IR. The PC4IR proposed eight recommendations:

- Invest in human capital;
- Establish an AI Institute;
- Establish a platform for advanced manufacturing and new materials;
- Secure and avail data to enable innovation;
- Incentivise future industries, platforms and applications of 4IR technologies;
- Build 4IR infrastructure;
- Review and amend or create policy and legislation; and
- Establish a 4IR Strategy Implementation Coordination Council in the Presidency.

These recommendations are outlined below.

Investing in human capital

The first recommendation is investing in human capital. This is only logical when you consider South Africa's rising unemployment rate, coupled with the brain drain we have seen in the last few years. Statistics show that unemployment in South Africa in 2020 is inching towards the 30% mark. This does not even include discouraged workers, who have given up looking for a job, which if included would put the estimate closer to 40%. Then, of course, there is the issue of the emigration of highly skilled workers, which Statistics South Africa (Stats SA) estimates to have been around 97 460 in the years 2006 to 2016 – a figure many have suggested is a gross under-representation. The solution is not only to foster growth to propel a stagnant economy but also for stakeholders in business, labour, government and civil society to create a coherent strategy.

This will not be an easy task, when you consider that the focus needs to be on both lowering the unemployment rate as well as preparing the current workforce for changes brought about by the 4IR. South Africa is one of the countries most vulnerable to 4IR disruptions. Consider, for example, the 2018 Human Capital Index (HCI) figures from the World Bank, which show that South Africa's HCI is just over 0.4 – the same as that of Benin and Malawi. However, South Africa has a higher GDP per capita than either of these countries and is categorised as upper-middle income, whereas Benin and Malawi are in the low-income group. What this means is that while South Africa has many economic opportunities, the majority of its citizens are not equipped to take full advantage of them.

In the 4IR, a mixture of skills stacked upon each other and aligned to industry would allow people to enter and leave the system, as part of a lifelong learning process. There ought, too, to be an investment in projects aimed at mass skills development. This investment can be scaled for exponential skills, pipeline development and market absorption, and it would be particularly effective in the manufacturing, agriculture and tourism sectors, which provide direct opportunities for such programmes.

As a matter of national culture, people and sectors from across society must be prepared to reskill and to approach upskilling as a continuous process. The education system at all levels must promote problem-solving skills, computational thinking, multi-disciplinary skills and systems-thinking, while equipping our students to master the social, economic and political worlds.

The government thus needs to prioritise a redesign of the human-capacity development ecosystem in order to link the entire pool of potential employees to productive and decent work, meaning that they deliver a fair income and workplace safety based on international labour policies. To achieve this, a comprehensive view of the entire human-capital system must be developed, and the leverage points that can be accelerated by the 4IR need to be identified. This will be facilitated at the Human Resource Development Council, assisted by the PC4IR and driven by the Digital Skills Forum, and it must include a timeframe on deliverable objectives.

The private sector, made up of both large businesses and Small, Medium and Micro Enterprises (SMMEs), ought to outline what skills are required and collaborate on strategic projects for mass skills development linked to various industries. Labour unions need to review their role in light of the 4IR and recommend appropriate worker protections. In order for this to be implemented, it has to be done in collaboration with the government.

Academic institutions – ranging from schools and universities to Technical and Vocational Education and Training (TVET) colleges – need to review their curricula, with a focus on the 4IR in order to ensure the relevance of qualifications based on requisite skills and the principle of lifelong learning. For instance, schools should promote digital literacy, while, at the same time, they should make a concerted effort to attract students to STEAM subjects. This, of course, cannot happen in silos but requires a concerted effort across the board.

The National AI Institute

The second recommendation of the PC4IR is to establish the National AI Institute. Again, this should be a collaboration between the public and private sectors – the rationale being that in South Africa, just as in many parts of the world, there is a greater capacity, especially around matters of AI, in the private sector than in the public sector. AI has three aspects, and the Institute must thus decide where it will invest its efforts.

The first aspect is the theory of AI, by which we mean the thorough study of AI, its architecture and the associated mathematics. This naturally includes the development of new AI methods. The second aspect of AI is the algorithmic part, which provides for coding. Fortunately, many companies, such as Google and Microsoft, have developed AI code that they provide for 'free'. Of course, nothing really comes for free, as the Chinese company Huawei realised when it discovered that the 'free' Android software from Google was no longer free when the interests of America and China clashed.

The third aspect is the application of AI. There are multiplicities of sectors and industries to which we can apply AI, among them manufacturing, agriculture, medicine and retail. The AI Institute will have to choose what areas of the economy it should invest in, in order to give South Africa a competitive economic advantage. It should also, of course, simultaneously co-create solutions with the rest of Africa.

The National AI Institute also needs to pay attention to the creation of AI solutions and apps, rather than only focusing on the theoretical aspects of AI. Of course, to create apps and solutions, one needs to be able to code. The AI Institute should, therefore, develop competencies in the area of integrating different software with different data sources in order to solve socio-economic problems. The Institute will need to be governed by a board or structures that have a fair representation of AI experts, as well as individuals from the public and private sectors across society. It should host experts from global centres of technological excellence, such as Silicon Valley in the US, Zhongguancun in China and Cambridge in the UK.

Furthermore, the AI Institute should work with the Deep Learning Indaba, which is developing AI expertise in Africa and is operational in 33 African countries. It should also work with initiatives such as Google Digital Skills for Africa as well as Zindi, a data-science community. The increase in Africa's population from 1.3 billion to 2 billion by the middle of this century presents a huge opportunity. When President Ramaphosa took over as the chair of the African Union (AU), he recognised the centrality of AI for Africa's economic growth. In consultation with the PC4IR, he announced the formation of the Africa AI Forum to exploit the emergence of AI opportunities in Africa. This has led to Google establishing the AI Lab in Ghana and Microsoft the AI Lab in Kenya. The AI Institute must be a conduit for AI knowledge to industry, society and government. It should also facilitate the expansion of AI know-how across the continent by drawing from both the local population and international expertise. It should use strategic partnerships in bodies such as the AU, the Southern African Development Community (SADC), the East African Community (EAC), the Economic Community of West African States (ECOWAS), Brazil, Russia, India, China and South Africa (BRICS), the US and the European Union (EU) to facilitate the development of people, expertise, skills and technology.

Advanced manufacturing

The third recommendation is to establish the Advanced Manufacturing Institute (AMI), which will focus on improving South Africa's competitiveness and take full advantage of the emerging technologies of the 4IR. The AMI should explore how to deepen automation so that South African companies remain competitive, and it should also look into how to automate industries, such as mining, so that they become safer and more productive by means of 4IR technologies. This automation should, for example, include volume measurement, which involves the use of drones and image-processing technology to measure the volume of stockpiles, such as coal, for instance. It can also be used in other areas of economic activity, including in the food and pharmaceutical industries. At UJ, a new technology called MinPET is in the process of being commercialised. MinPET provides real-time 3D imaging of locked diamonds (diamonds that are embedded in kimberlite rocks) and, as the system is fed with more data, the results improve exponentially. Another way we could use intelligent automation is by utilising self-driving cars in underground mines, especially in areas that are dangerous for human beings. Sensors can be used in conjunction with AI to monitor all aspects of operations in the mining as well as manufacturing industries. In a decentralised fashion, the AMI should develop principles applicable to all types of industries, such as the automotive industry, the fast-moving consumer goods (FMCG) industry and the aerospace industry. The AMI should also explore how the limited spectrum required by the telecommunications industry – the radio-frequency bands allocated to companies and other entities that allow for communications and enable technologies such as 5G – could be allocated to companies so that they can use technologies such as the IoT to improve their productivity. The AMI will need to be driven by a board with the necessary technical expertise in the broad range of industries, and it should be linked to agencies – such as the Council for Scientific and Industrial Research (CSIR) – that boast deep technical skills in various manufacturing sectors. It should establish ties with institutions of higher learning that have expertise in manufacturing. It should also build relationships with international centres of excellence in manufacturing in countries such as Germany, China, Japan and South Korea, and, at the same time, it should coordinate ongoing local conversations between government, academia and industries.

Data capability

The biggest driver of the 4IR is data, often referred to as 'the new oil'. Whoever controls the data controls the destiny of people. In this case, what we mean by data is the units of information that can be collected, such as personal or demographic data, for example, and used by machines to make intelligent decisions. Without data, AI is little more than a computer code without consequence. The general rule for AI is that the more data that is presented to it, the more effective it is. Countries that gather more data – such as China and the US – will dominate the economic, political and social worlds.

On 19 February 2020, the EU, which has lagged behind China and the US in the data-driven digital economy, published the *European Strategy on Data*. This proposes the establishment of a single European data space, thus ushering a single data market to facilitate data gathering, storage, processing and use, and to promote the development of data-driven companies in the EU. The AU should follow suit in creating a single African market for data to foster an ecosystem that will allow the development of data-driven companies on the continent.

Allow me to create a fictional scenario. Let's imagine that Makatu needs to go to Tshilidzini Hospital in Limpopo. As part of his medical examination, he is required to have a CT scan. Later, Makatu is transferred to Garankuwa Hospital, which is more than 400 kilometres away, where he has to have the same scan. This is because there is no practical transfer of data from one public hospital to another in South Africa. Now, let's suppose Thendo wants to automate the process of reading medical images produced from scans in order to diagnose tuberculosis and pulmonary embolism; Thendo will need these medical images of all the patients who have visited public hospitals, but there is no single databank that houses all scans taken at government facilities. This simply means that the AI system that Thendo builds will not be as effective as the one built by Bo Xing from Tianjin, because China's medical records are integrated. There is, therefore, a need to consolidate data capabilities with the South African state.

The fourth recommendation of the PC4IR is to secure and avail data to enable innovation – a factor that is critical for building e-government services across sectors, such as health, transport and justice. This could be achieved through the creation of the National Data Centre, which would consolidate the available computational power and create a national data centre that will become the countrywide repository for all our data, including that of health. This can be done alongside existing data companies. However, it is important that cybersecurity is bolstered in order to safeguard the public. Already, the South African government has COMSEC – a cybersecurity company linked to the National Intelligence Agency – which was established in 2003 in order to secure the government's communications against any unauthorised access and also from technical, electronic or any other related threats. To be competitive in the 4IR, the government would need to strengthen COMSEC's capabilities to include cybersecurity. A cross-departmental Chief Data Officer could facilitate this. The private sector could engage with the government on critical datasets required for innovation and service-delivery collaboration and perhaps share data sets with the government. Academic institutions, of course, will be an essential pillar on data best practices and ethical data sharing.

Incentives

Steven Landsburg was once asked to summarise the whole of economics in one word, and his answer was 'incentives'. Rational people are inherently self-interested. For them to participate in the economy, they need incentives. South Africa is experiencing de-industrialisation, and the reason is simply the decline in competitiveness. South African companies facing an expensive and weak or unskilled labour force have not sufficiently invested in technologies, particularly those of the 4IR. Given this lack of investment in technology, the fifth recommendation of the PC4IR is to incentivise future industries, platforms and applications of 4IR. This means that companies will need to be incentivised to use 4IR technologies in order to improve South African competitiveness. These incentives should include tax breaks and support for research and development using organisations such as the CSIR, the National Research Foundation (NRF), the Medical Research Council and the Agricultural Research Council, thus supporting the acquisition and application of advanced technologies in the manufacturing of goods and delivery of services. Part of this exercise will include providing additional support for the development of new SMMEs, and it will also include growing existing businesses in the 4IR space to create solutions that address South Africa's development challenges. This requires improving the 'ease of doing', including tasks such as registering a patent, reducing the cost of 4IR enterprises with regards to customs and taxes, and enabling global competitiveness and expansion. The government will thus need to establish appropriate incentives, remaining cognisant of the country's precarious fiscal position.

Infrastructure

The sixth recommendation is to build 4IR infrastructure to integrate with existing economic and social infrastructure. To create a coherent and comprehensive infrastructure network, we need to look at the generation and delivery of energy, the extension and improvement of water, health and educational infrastructures. The first step would be for the government to develop a comprehensive set of priorities with achievable timelines. In 2012, the government adopted the National Infrastructure Plan in order to create jobs and strengthen the delivery of essential services. However, much of this needs to be done with urgency because infrastructure is integral to the 4IR. Businesses can play a vital role in engaging with the government on such projects that can be tested and scaled. At the same time, labour unions can facilitate solving negotiation deadlocks that prevent timeous project rollouts.

Legislation

A few years ago, I visited Thohoyandou, a small town in Limpopo, close to the village where I was born. Because my family of five could not be accommodated at my parents' house, we decided to rent a place through Airbnb. This transaction was done via the Airbnb app and, quite coincidentally, the owner of the house happened to be someone I actually knew – meaning that this transaction could have been completed without Airbnb. Airbnb is domiciled in the US, and yet it is doing business in South Africa. And even though Airbnb is operating here, it is not subjected to the South African tax system. The reason for this is that the South African legal system is not yet adapted to developments in the 4IR.

Given the mismatch between our legal system and developments in the 4IR, the seventh recommendation of the PC4IR is to review, amend or create policy and legislation. To ensure our legal system is in line with the 4IR, parliament will need to look at all our legislation and update it in line with the 4IR. This would require members of parliament and cabinet to become 4IR and science literate. In particular, the generation of IP rights stands out in this context because the very principle of a creative and knowledge economy implies the rapid production of new technologies, artefacts and processes for commercialisation and scale. This will need to include reviewing our tax laws so that they bring platform companies, such as Uber and Airbnb, into our tax regime.

Implementation

There is silent acknowledgement in South Africa that the government talks too much and implements too little. The country should move away from being a nation of talkers to one of doers. Of course, implementation requires capacity and resources. To move South Africa from just talking to doing, the PC4IR recommended the establishment of the 4IR Strategy Implementation Coordination Council in the Presidency. This council would coordinate government departments responsible for 4IR-related programmes, and it would also coordinate initiatives across the public and private sectors, labour and academia. This would require resourcing and budget allocation aligned to the mandate to ensure that there is a single point of coordination with government departments for the council.

PART 1

Industry

This section looks at the impact the technologies of the 4IR have had on industrial sectors, such as mining and electricity. By way of introduction, it begins with an explanation of the importance of creating safe structures by reducing errors and focusing on efficiency through AI, and these are then applied to specific examples. The takeaway is that AI increases efficiency and accuracy in that it mitigates any uncertainties that may arise, and it is also able to solve complex problems. Across the various industrial sectors, AI speeds up decision-making, reduces error rates and increases efficiency, ultimately ensuring that the accuracy of 4IR technology is unparalleled.

CHAPTER 4

Safety of structures

My grandmother, Vho-Tshianeo Marwala, was my first engineering teacher. She did not go to Case Western Reserve University in the US to study Mechanical Engineering as I did. She did not go to the University of Cambridge in England to study for a doctorate in AI as I did. She never left South Africa. All her life was spent primarily within a 25-kilometre radius of the village of Duthuni in Venda. And yet, despite all these limitations, Vho-Tshianeo was an excellent teacher who taught me engineering from an early age. She taught me how to predict the collapse of building structures. She would often look at the sky and predict whether it would rain, based on the orientation of the clouds.

In April 2018, three children were killed, others rushed to a hospital, when an abandoned building collapsed in Johannesburg. Herman Mashaba, the city's mayor at the time, stated that the tragedy could have been avoided. In March 2018, three people died when a building in Durban collapsed. Again, this tragedy could have been prevented. In 2015, temporary works that were part of a pedestrian and cyclist bridge collapsed near the Grayston Drive offramp on the M1 highway in Johannesburg, killing two and injuring many others. How then do we prevent buildings and bridges from collapsing? Once the background of the 4IR is understood, its real applications can be teased out. For instance, can we predict and prevent the collapse of buildings or bridges? In seeking a solution, I do not have to look beyond the lesson in engineering that my grandmother taught me.

Clay pots

Vho-Tshianeo used to make clay pots (see Figure 4.1) with impressive dexterity. The art of making clay pots is rich in lessons in engineering: the mechanics of supply-chain management, metallurgy, applied mathematics, thermodynamics and AI. Studying how to make clay pots can guide us in

Figure 4.1 *Clay pots similar to the ones crafted by Vho-Tshianeo*

knowing how to prevent buildings from collapsing. Firstly, one needs to identify a source of good clay, and this requires knowledge of materials science. Then the clay is delivered to a manufacturing plant where it is processed and formed into pots. This requires 3D visualisation; the ability to form shapes in one's mind. Then the pots are put in the sun so that they can dry. After they have dried, a furnace is created where these pots are baked so hot that they look red because of fire. This requires knowledge of thermodynamics, which is the relationship between heat and other forms of energy. Then the fire is allowed to extinguish, and the pots are slowly cooled. Slowly cooling the pots is called annealing, a process learned in metallurgical engineering, which focuses on the physical and chemical behaviour of metallic elements.

If the pots are cooled too fast, they crack. The process of annealing is such a powerful concept that an AI algorithm called 'simulated annealing' was created based on its principles. This looks at the probability of various solutions to find the optimal one. Simulated annealing is used to find the shortest distance between two locations. Technologies such as Google Maps and global positioning system (GPS) in our cars use simulated annealing.

The Boltzmann equation and the organic intellectual

Understanding simulated annealing requires being familiar with the Boltzmann equation, which was devised by the Austrian scientist Ludwig

Boltzmann. My grandmother was, however, able to understand and appreciate the annealing process without any knowledge of the Boltzmann equation – she simply followed the practicalities of annealing. This notion of knowing a concept without understanding the theoretical point of it (that is, the Boltzmann equation) is what the Italian philosopher Antonio Gramsci called organic intellectualism.

The concept of organic intellectualism should not be understood to be suggesting that education is not essential, but rather that education is vital in its totality. In other words, we should incorporate indigenous knowledge systems into our curriculum, research and innovation agenda. It means that, as Africans, we should not alienate ourselves from science. For example, the *Hoodia gordonii* plant, used in South Africa for hundreds of years by the Khoisan, suppresses hunger. This makes it a perfect weight-loss remedy. The ingredient P57 was isolated from the plant and patented by the CSIR and was subsequently sold to Pfizer. It is evident from this example that our ancestors understood science, especially its practical implications.

Predicting the failure of structures

It is the last stage of making clay pots that is instrumental in allowing us to predict the failure of structures such as buildings and bridges. My grandmother took each clay pot, tapped it and listened to the ringing sound. Based on what she heard, she was able to tell whether the pot was good or bad quality. If the pot rang for a long time, it meant it was baked well. When it rings for a long time, mechanical engineers would call it a 'lightly damped structure'. If it rings for a short time, it is a bad-quality pot, and mechanical engineers call this a 'damped structure'. My grandmother was not a mechanical engineer and, therefore, she had no understanding of the theory of damping. Yet, she understood the principles of damping and its implications on the structural integrity of pots.

Depending on how it rings, one can predict whether there are pockets of air trapped inside the walls of the pot. This process of using sound to test whether the pot is good or bad quality is what engineers call 'non-destructive testing'. This procedure is routinely applied in aerospace engineering to assess whether aeroplanes have cracks on their bodies. The test, as taught to me by my grandmother, shows us that 'everything has something to say; all you need to do is to know how to listen to it'. In effect, what she was saying was that the clay pot can tell us through sound after being tapped whether it is of good quality or not.

Vibration data

In my book, *Condition Monitoring Using Computational Intelligence Methods*, I take the concept of listening to objects, in the same way that my grandmother listened to her pots, into the 4IR.[1] The sound that my grandmother was listening to is what we in engineering call 'vibration data'. This vibration data is processed using the Fourier analysis, a tool that breaks the data into a series of cycles and identifies their natural frequencies. In the case of using the Fourier analysis to determine the integrity of a clay pot, for example, we would first record the frequencies at which the clay pot vibrates after being tapped. The Fourier analysis identifies many component or constituent frequencies of a clay pot. This would help us identify particular frequencies that would alert us to the fact that the pot is damaged. The study of the data, represented in cycles, does not need to be done by a human brain – in fact, in the 4IR it is performed by machines. Universities offer courses in vibration analysis, signals and systems, thermodynamics, as well as AI, which are all necessary to take the framework my grandmother taught me into the 4IR.

Artificial brain

Due to her deteriorating hearing, the older Vho-Tshianeo became, the more she disposed of good pots. AI machines do not suffer from hearing loss. A clay pot diagnostic device is considered smart if it can analyse information and extract insights beyond the obvious. While computers traditionally relied on people to tell them what to do and how to react, AI machines can learn and make decisions on their own. Just as we would train a dog to do tricks, we can instruct AI to carry out certain tasks.

We may remember the movie *WALL-E* that was released more than a decade ago. WALL-E is a garbage-cleaning robot stuck on Earth, who has learned to repair himself with harvested parts and to collect souvenirs while going about his daily work. He meets EVE, a robot charged with finding organic matter on Earth. She is from a spaceship that is home to the remaining humans, who are all pampered by robots that do specific tasks, such as sweeping the floors or piloting the ship. This is the idea behind AI: systems that can do many menial jobs for us, making us more efficient and freeing up our time.

These kinds of robots could well become a reality, but are certainly not as advanced quite yet. The Roomba, for instance, is a robotic vacuum cleaner. It can scan the size of a room, identify obstacles and remember how to clean the carpet and which routes work best.

Personal assistants such as Siri or Alexa are able to recognise your speech or, in other words, understand what you want or need, analyse the information to which they have access, and provide an answer or solution. Many of us may use Siri to answer a question quickly. If one asks Siri to list the latest movies on Netflix, she will scour the internet and come back with an answer. In fact, Siri also has a few personality traits programmed into her, so engaging with her can feel as though you are having a conversation. These are by no means deep conversations, but she will indeed reply. One can ask Siri to tell a story and, around Halloween, Siri can give costume ideas. If one asks, 'What's the meaning of life?' I have it on competent authority that one of her responses is that all evidence to date suggests that it is chocolate. Assistants such as Siri and Alexa, much like other software that uses AI, continuously learn about their users until they can accurately anticipate your personal needs.

So, AI is advanced and can be used to listen, for instance, to the sound emanating from a clay pot, as shown in Figure 4.2. Here, the AI algorithm takes the sound information coming from the clay pot, processes it and decides whether the clay pot is good or bad quality. In a way, the AI systems replace my grandmother's brain. The advantage of these systems is that, unlike my grandmother, they neither tire nor experience mood swings that affect decision-making capacity.

The levels of damage-detection in clay pots

The framework proposed earlier decides whether the clay pot is good or bad quality. There are three levels: the presence, location and intensity of the damage. The sector in which it is essential to know the levels of damage is manufacturing. Suppose that in a car-manufacturing plant there is a

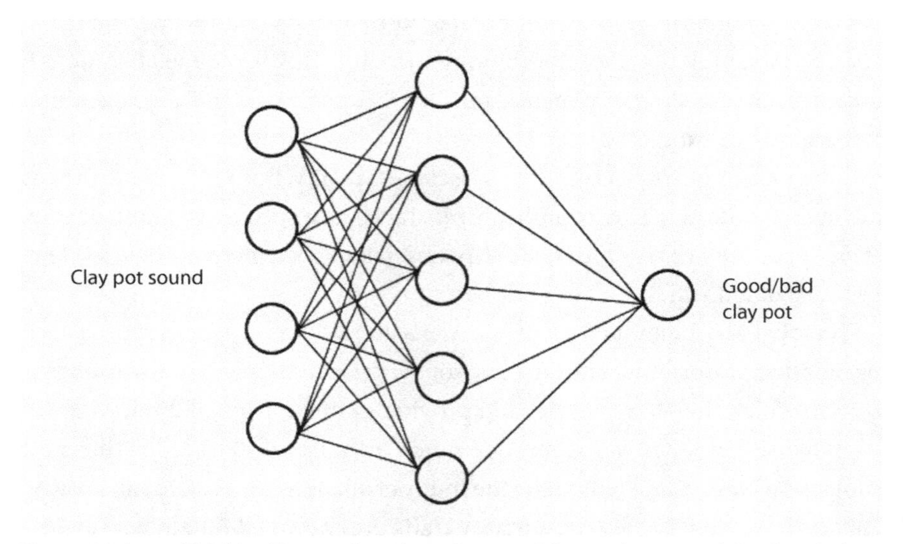

Figure 4.2 *An artificial intelligence algorithm takes sound from a clay pot and decides whether the pot is good or not*

system that manufactures car doors and monitors whether those doors are of suitable condition. Knowing whether the door is in good or bad condition will only prevent us from installing a substandard door, avoiding a situation that will inevitably lead to customer complaints and thus devalue the manufactured product. It, however, does not tell us where the damage of the car door is located, nor does it tell us the extent of the damage – information that is essential to fixing the root cause of the problem. The first level of damage-detection is thus the identification of the presence of damage, also called Type 1. The second level of damage-detection is the location of the damage, called Type 2. The third level is the extent of the damage, called Type 3.

When my grandmother was evaluating damage in clay pots using her ear to listen to the sound emanating from them and her brain to process the sound and make a decision, she was evaluating Type-1 damage-detection. AI systems can assess all these levels of damages; all we need to do is to replace the good/bad clay pot in Figure 4.2 with the location of damage for Type-2 damage-detection or the extent of the damage for Type-3 damage-detection.

Other applications

This framework of assessing the structural integrity of clay pots can clearly be used to monitor the safety of buildings and bridges, and people's health, for example. In this regard, data-acquisition devices or sensors are embedded in buildings, bridges or even people's bodies, and the data gathered is relayed to an AI machine. This machine analyses the data and makes a decision as to whether the building or the bridge is structurally weak, or the person is at a high health risk. In the case of imminent danger, automated messages can be relayed to allow relevant measures to be sought. This allows for the buildings or bridges to be secured before they collapse, thereby saving lives. It also allows a doctor to be called to save the life of a patient. For example, in my book *Artificial Intelligence Techniques for Rational Decision Making*, I detail how I used this technology to analyse and predict epileptic seizures before they occur and then allow for intervention.[2] This framework has also been used in medical diagnostics whereby images of lungs are analysed to determine whether the patient has a pulmonary embolism – in much the same way that it is used to predict leukaemia. Similarly, this can be used in electricity distribution or asset management.

AI technology has already proved to be an efficient alternative approach to classic modelling techniques. In contrast to conventional methods, AI can deal with any uncertainties that may arise and is useful in helping to solve complex problems. Ultimately, this cuts down on the tedious aspects of engineering by making the process of decision-making faster, reducing error rates and increasing efficiency. When we are dealing with people's lives, the accuracy of 4IR technology is unparalleled.

CHAPTER 5

Aerospace

On a summer evening in 1999, John F Kennedy Jr (son of assassinated US President John F Kennedy), his wife, Carolyn, and his sister-in-law, Lauren, boarded an aircraft in New Jersey for a flight to Massachusetts to attend the wedding of Kennedy Jr's cousin, Rory. The three never arrived at Martha's Vineyard, and the wedding did not take place as scheduled. The story is that the pilot, Kennedy Jr, lost control of the aircraft, and it crashed into the Atlantic Ocean fifteen minutes before arrival. There are two possible reasons why Kennedy might have lost control of the plane. Either the gyroscope – the device used to determine whether aircraft are going up or down or moving forwards or backwards – stopped functioning, or Kennedy stopped relying on it. Either way, once a pilot no longer receives information from the gyroscope, he or she can become disoriented, and the plane eventually does what is called a 'dead man's dive' and crashes. When gyroscopes fail, a dead man's dive is almost always inevitable. Aeroplane controls are delicate and must be handled with due care.

In one of the deadliest aviation accidents in recent times, an Ethiopian Airlines Boeing 737 Max 8 crashed in 2019, a mere six minutes after taking off, killing all 157 people on board. Just three minutes after departure, the captain had requested to return the plane to the runway. Six months before, in 2018, a Lion Air Boeing 737 Max 8 had crashed in Indonesia, thirteen minutes after taking off, killing 189 people. According to the theory of probability, when something happens once, we regard it as an accident, but when the same thing happens twice, it is probably because of a systemic issue. These two new Boeing 737 Max 8 aircraft crashed within six months of each other, so it seems that this family of aircraft has a systemic problem. And it hasn't gone unnoticed. In 2019, Boeing lost orders for 87 commercial aeroplanes – the first time in at least three decades the manufacturer lost orders for the year. In December 2019, Boeing failed to log any 737 Max orders.

International reaction

Within 24 hours of the 2019 crash, China became the first country to ground all Boeing 737 Max aeroplanes. Other countries followed suit, including those of the EU as well as Canada, Japan and South Africa. What is interesting is that the US did not rush to ban the aircraft from the skies. Was this because of American patriotism or because the American regulators were pandering to the public? Boeing is an American company that is closely linked to the political establishment of the US – in 2018, for instance, Boeing allegedly spent US$15 million on political lobbying, a practice in which private individuals or companies attempt to influence elected politicians to act in a particular way that is favourable to them. Boeing is the largest supplier of aerospace goods and services to the US government. It is also the maker of Air Force One, the aircraft used by the president of the US. The fact that the US was one of the last countries to ban this aircraft is an example of the possible toxic confluence of politics, money and lobbying.

Closer to home, the Rooivalk, an attack helicopter, was developed in the 1980s as part of South Africa's military arsenal during the Border War. The project was marred by delays and escalating costs that made it unfeasible to use during the war, so the hope was that it could be sold to foreign militaries in the 1990s. However, this did not materialise, as Malaysia, for one, opted for the US Apache instead, and South Africa failed to win a Turkish tender. This was largely due to the US's political influence and the ongoing arms-deal scandal in South Africa, which left the South African National Defence Force as manufacturer Denel's sole Rooivalk-helicopter customer.

How do aircraft fly?

In order to make sense of what might have caused the Boeing 737 Max aeroplane crashes, and then to identify ways that 4IR technologies could help us avoid such disasters, it will be useful for us to have a basic understanding of how aircraft fly.

When I was a Mechanical Engineering student at Case Western Reserve University, I was an active member of Students for Exploration and Development of Space. Later, I worked in the area of damage detection, which is of particular interest in the aerospace community. When I became a researcher, I published papers in the *American Institute of Aeronautics and*

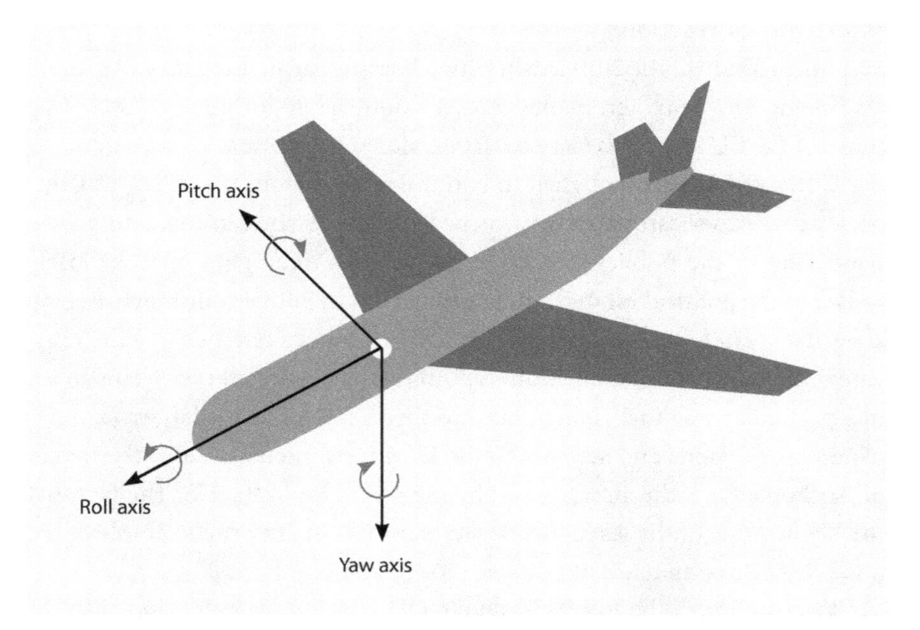

Figure 5.1 *Aircraft manoeuvres: pitch, roll and yaw*

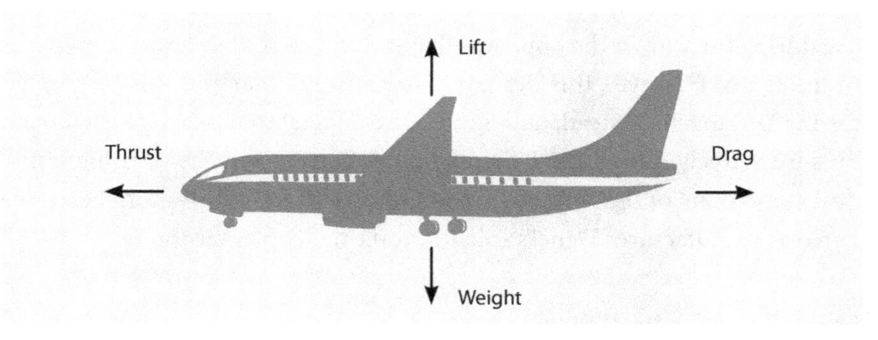

Figure 5.2 *Forces of flight: weight, lift, thrust and drag*

Astronautics (AIAA) Journal as well as in AIAA's *Journal of Aircraft*.

Now, how do aircraft fly? Firstly, the fuel that is burned is directed in such a way that it releases a high-speed gas that gives the aircraft the propulsion it requires to move forward. Secondly, the pilot (or autopilot) is able to make the aircraft take off, move around in the air, and land by orientating the aircraft's wings, as well as by altering its vertical position in the air, in varying ways, in order to produce different manoeuvres.

In flight mechanics, aircraft make three types of principal manoeuvres:

the pitch, the roll and the yaw. These are shown in Figure 5.1. Pitch is when the nose of the aircraft is either pointing down, forward or up. If, for example, the plane makes a straight pitch, it moves forward. If it makes a down pitch, it turns down, and if it makes an up pitch, it goes up. The roll manoeuvre is when the body of an aircraft rolls. If the aircraft rolls out of control, then it loses its wings. The yaw move is when the aircraft spins horizontally. If it spins out of control, it becomes unstable, and will likely crash. These three manoeuvres are necessary for the aircraft to fly, turn and land.

While the aircraft is flying, it is subjected to a number of forces – shown in Figure 5.2 – including the weight of the aircraft. It is the weight of the plane that tends to make the aircraft fall. The second force, which counterbalances the weight, is the lift force. The third force is the thrust, which moves the aircraft forward, and is a direct result of the propulsion resulting from the high-speed emission of steam from the combustion engine. The combustion engine is where the fuel is burned in order to produce propulsion.

The final force is the drag. As the aircraft moves through the air, it experiences a drag force. Drag force is much the same as we experience when trying to run in water in a swimming pool, for example. The reason why we are unable to run fast is because of the drag.

What went wrong with the Boeing 737 Max 8?

It seems that the Boeing 737 Max 8 had a problem with the pitch manoeuvre; the aircraft was unable to determine whether it was climbing, descending or heading straight. Several individuals had – before the most recent crash – identified the technical problem with this aircraft. This flaw had to do with automation.

In general, there are three ways in which automation can happen. The first is a case in which a machine cannot make decisions without the involvement of a human being. The phrase 'human in the loop' here essentially means that both human and machine make the decisions jointly. The second way is 'human on the loop'. This is the case when a machine makes all the decisions but with a human watching and only intervening if there is a problem. The third way is 'human out of the loop'. This is when a machine makes all the decisions, with no human present. The autopilot system is an example of the second way described above. The automation system makes all the decisions, but a pilot can intervene if he or she is not satisfied with the action taken by the machine. A pilot would, for instance, have to switch off the autopilot when an aircraft is pitching down when it is supposed to be climbing. The problem with the Boeing 737 Max 8 is that its flight-control system is not stable. What, then, is this flight-control system?

What is a control system?

A control system is a device used to control a system such as an aircraft. There are two types: open-loop and closed-loop systems. An open-loop system does not have a feedback mechanism, one example being a washing machine. All you do is put the laundry in and set the parameters, such as whether the laundry is cotton, and the machine will do the rest. It will not determine whether the clothes are washed well. The open-loop system is illustrated in Figure 5.3.

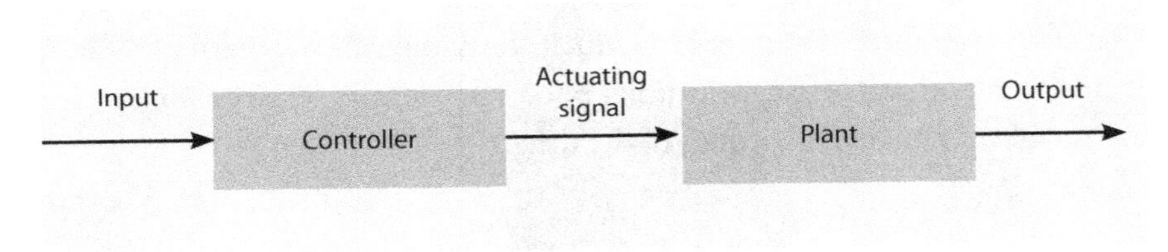

Figure 5.3 *An open-loop system*

The second type of control system is a closed-loop system, as shown in Figure 5.4.

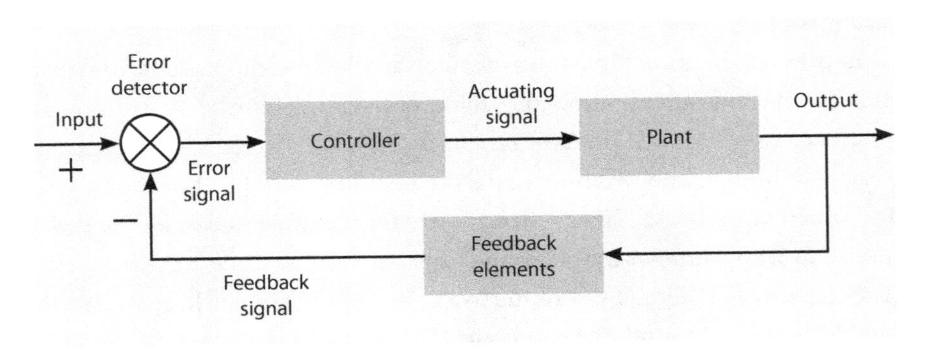

Figure 5.4 *A closed-loop system*

For example, if a pilot wants to climb to an altitude of 800 metres above sea level, then the system will make this happen by determining the speed, pitch, roll and yaw configuration schedules, for instance. Therefore, for the objective of reaching an altitude of 800 metres to be achieved, the system requires information such as the present altitude, speed and pitch, among others. This is because the aircraft is unable to climb to a particular altitude until it knows its current elevation. We call the concept of understanding the current state, altitude, speed, etcetera, in order to be able to control the aircraft, a 'feedback mechanism'. Thus, the control system consists of a data-acquisition device, a model that makes sense of the measured data, and a feedback mechanism. The feedback mechanism serves to ensure that the path the control model recommends to reach a desired goal or objective is still valid.

The mathematical and physics models of different aircraft in flight

For the control system to work, we should have an accurate model of the aircraft – in other words, a set of mathematical equations that are able to tell us about the behaviour of the aircraft in flight. The model should predict, as accurately as possible, how the plane will be impacted by different forces of flight and therefore how it should be manoeuvred precisely in order to fly it in the desired way and to avoid crashes. One of the leading and most charismatic scientists of the 20th century, Richard Feynman, once said that if a model does not predict reality, then it is wrong and needs to be discarded. In aerodynamics, the models used are based on what is called the Navier–Stokes equations. These describe the physics of engineering phenomena and can be used to model water flow in a pipe or airflow around a wing, for instance. However, the problem with the Navier–Stokes equations is that they cannot be solved. As a result, designing an aircraft is a tedious and expensive process that involves trial and error.

How do we fix what went wrong with the Boeing 737 Max 8?

There are a few reasons why aircraft are not able to determine the pitch correctly. One of these is that the sensors used to measure the pitch might have a factory flaw. Another reason why control systems fail is instability. In this regard, the control system should be designed so that it avoids operating regimes that will make it unstable.

So, you may ask, what do the fate of the Boeing 737 Max 8 and the failure of its control system have to do with the 4IR? The control system is an essential component of automation, and it is thus a principal driver of the 4IR. In the third industrial revolution, control systems were not as intelligent as they are in the 4IR, because they were based on fixed rules. Due to this, automation was inflexible, and thus human beings were needed to constantly oversee and reconfigure the work of machines. The 4IR, on the other hand, is driven by intelligent technologies such as AI and machine learning. Thus, automated control systems of the 4IR are smart. However, this intelligence makes them vulnerable to stability problems. Researchers still need to develop robust methods that will make these intelligent systems stable in order to avoid disasters, such as those seen in the Boeing 737 Max 8 crashes. By using advances in AI, it is essential that experts design these systems so that they are able to oversee their own actions and decisions in order to ensure they are fail-safe.

When an aircraft, such as the Boeing 737 Max 8, crashes, its black box, which records flight information, must be retrieved and the data must be analysed in order to understand what caused the crash. It is reported that the black box recovered from the Ethiopian Airlines crash was taken to France to be decoded. Ethiopia lacked the decoding software, and other countries, such as Germany, similarly lacked it. However, the African continent, through organisations such as the AU, should develop such software as a home-grown solution. This will deal with the historical trend of African countries looking to Europe for answers. This brings forth the question on what African universities are doing to create a cadre of graduates who can understand sophisticated technologies, such as automation, AI and control systems.

Despite the failure of the control system in the Boeing 737 Max 8, autopilot remains an essential tool typical of the 4IR that helps us ensure our safety. For example, it can help in situations where the pilot is fatigued or is faced with landing the plane in difficult conditions. However, we need to ensure that we use AI optimally to improve such systems.

Moving forward

There are lessons to be learned from the aftermath of the Boeing 737 Max 8 crashes. Moving forward, it is clear, firstly, that we need to study and understand the complexity of the 4IR, including its downsides. Secondly, we need to guard against the toxic confluence of politics, economics and lobbying to ensure our safety – even at the expense of profit maximisation. Thirdly, we need to capacitate our universities so that they have the necessary resources to teach complex and evolving subjects, such as control systems. African universities should create a cadre of graduates who can understand these sophisticated technologies, such as automation, AI and control systems. Any alternative is unthinkable.

CHAPTER 6

Mining

On 26 June 2018, South Africans woke up to the news that another person had died at Sibanye-Stillwater gold mine, bringing the total number of people who had died at the mine to 21, accounting for nearly half of all fatalities in South African mines in that year to date. According to the US Department of Labor, the annual number of deaths in coal mines in that country dropped from 3 342 in 1908 to 15 in 2017. In South Africa, mining fatalities dropped from 290 in 2000 to 82 in 2017. The sad reality is that despite improvements, our mines have not achieved the safety levels that South African miners deserve. Moreover, people at the coalface of mining are still underpaid, earning less than R10 000 per month.

This is especially damning when you consider that mining has historically been one of the driving forces behind South Africa's economy. Yet, the industry has been in a downward spiral for many years. The contribution of the sector to the country's GDP has declined over the past two decades, from a high of 21% in the 1980s, when much of the economy depended on the industry, to around 8% of GDP. Yet, the sector makes up about a third of the market capitalisation of the Johannesburg Stock Exchange (JSE). According to Minerals Council South Africa, the mining industry constitutes 8% to 15% of South Africa's total employment, contributes 15% to foreign direct investment (FDI) and accounts for 20% of private investment. The sector also sits on extensive reserves and is a world leader in the production of platinum, vanadium, vermiculite, manganese and chromium. Yet, according to the Fraser Institute's 2018 *Annual Survey of Mining Companies*, 'South Africa is the tenth-worst country in the world in which to own mining interests, out of 91 with significant mining industries.'[1]

As the 4IR takes hold, we need to interrogate ways in which we can make industries safer. How do we prevent these casualties? Then we need to look at how we resuscitate a dwindling sector, given how heavily our economy still relies on it. How do we bring an industry grappling with the first, second and third industrial revolutions into the 4IR? How do we infuse the technologies of the third and the fourth industrial revolutions – such as automation, AI, blockchain and the IoT – into the mining industry?

Working in mines is dangerous

Growing up in Venda in Limpopo, I witnessed first-hand how one of the ways in which parents motivated their children to study hard was to threaten that if they did not study, they would end up working 'mugodini' – in other words, 'inside a hole'. By this, they meant working in the mine. Given the fatalities tallied by the industry every year, it seems this threat was effective, because very few people from Venda ended up working in the mines. The reality many were thus avoiding, which remains to this day, is that, despite technological advancements and legislation that has made mining safer than in the past, miners still die from explosions, cave-ins and equipment failures. Added to that, they are also exposed to fatal and chronic conditions that persist long after they have retired.

In order to fully grasp the dangers of working in the mines, it is imperative to understand how mines work. There are different types of mines, but the two that deserve our attention are open-cast and underground mines. Open-cast mines, also called open-pit mines, are where you dig for materials on or near the surface level of the earth. This involves digging a pit in the ground, from which minerals are extracted. Underground mines are those with tunnels under the surface of the earth that allow you to access minerals. South Africa has one of the deepest mines in the world: the Mponeng gold mine, which is nearly four kilometres deep. Ominously, the term 'mponeng' means 'come and see me'.

Because most underground mines are very deep, the environment is extremely hostile. Firstly, they are very hot, with temperatures rising to 55 °C. Secondly, the conditions are very humid, which means that people have difficulty breathing, and this, in turn, has a long-term effect on their health. Thirdly, underground mines are very noisy, simply because to dig for minerals, one uses hydraulic drills. Because of such high noise levels, which can be louder than 80 decibels, mineworkers progressively lose their hearing. In fact, I'd be so bold as to say that mines are so dangerous that it is, in many ways, irresponsible to send human beings down there to work, especially in the era of the 4IR. In the 4IR, machines, by means of robots, for example, can execute any number of tasks that were originally designed for humans – although, of course, using technologies of the 4IR in mines also comes with the challenges of high temperatures, humidity and a generally hostile environment where there is plenty of dust.

The political economy of mining

If we are to measure an industry's citizenship in the 4IR by tallying the number of semiconductor devices – that is, phones, computers, intelligent robots – that are used in that industry, then mining is lagging compared to, for example, the automotive, defence and manufacturing sectors. Examples of semiconductors used in the mining industry include metal oxide semiconductor sensors that monitor air-contamination levels, and semiconductor gas sensors that are used for detecting fires underground.

But to understand the slow pace of the advancement to the 4IR by the mining industry, we also need to understand its political economy. Mining companies acquire mining rights from the Department of Mineral Resources. A mining licence grants a right to mine a particular mineral in a specific location over a particular period. To get this licence, the mining company must do an environmental-impact assessment as well as consult with the local community. These companies then put mining infrastructure in place, employ workers and operate the mine. The agreement that government reaches with these companies is that they will hire people, pay taxes and pay royalties to the government in exchange for the right to mine.

In game theory – a mathematical field popularised by the movie *A Beautiful Mind* – there is a concept known as 'zero-sum game'. This means that any gain by one party is always a loss to another party. The mining industry is a zero-sum game, because the gain of minerals by these companies is necessarily a loss to the minerals balance sheet of a country and, in many ways, a loss for the climate-change balance sheet. The minerals balance sheet reflects the aggregation of all mineral assets in the country. In contrast, the climate-change balance sheet gives an indication, among other things, of the level of toxic fumes in the environment, reflecting the country's environmental health. What government hopes is that employment, taxes and royalties compensate for the loss of mineral resources and environmental health. Another reality is that employment relations in the mining industry have not changed much in the last century – wages remain low, and the work environment is not yet 100% safe.

Slow move to the 4IR

Now, why has the mining industry not invested in the technologies of the 4IR?

As we have seen, the mining environment, especially when it comes to deep mines, is hostile to these technologies. The high temperatures, humidity and noise levels mean that semiconductor devices, such as an AI-powered robot, do not last.

There are four main reasons why the underground environment is unsuitable for the implementation of 4IR technologies.

The first is the high temperatures underground. One element that is crucial to the effective functioning of semiconductor-based technologies is temperature control: computers have to be cooled down continuously by internal fans, otherwise they will overheat or even melt. This is why, when you operate your laptop while it rests on your lap, it becomes hot – even though the computer has sophisticated fans that cool it. Data centres thus consume large amounts of energy because they have to be cooled, and the most significant contributor to the cost of data centres is electricity. Electricity is used to store and process information and to cool emissions created when data is stored and processed. This is also why in South Africa, where Eskom is struggling to satisfy the demand for electricity, it is challenging to invest in data centres. When we put such semiconductor devices in mines, where temperatures can rise to 55 °C, the amount of energy needed to cool them is high, and the relevant technology is not yet available.

The second reason is humidity. In Metallurgical Engineering, we learn that moisture in the air (that is, humidity) causes materials to rust quickly. We also learn that humidity makes devices unpredictable because humidity levels influence the performance of machines.

The third reason is that the environment underground is characterised by uneven surfaces, with obstructions such as piles of debris or rock formations. Therefore, a robot that has been designed to operate on a factory floor is unable to work optimally underground. This is because considerably more resources have been invested in factory robots than in underground robots.

The fourth reason is that the sensors needed to power these robots are designed for above-the-surface environments rather than underground environments.

Moving mining into the 4IR

Given all these limitations, how do we reconcile mining and the 4IR? Given that government grants companies the right to mine in return for employing people and paying taxes and royalties, how do we handle the fact that the 4IR will decimate employment? How do we reconfigure taxation, given the fact that the mining industry is an extractive sector? How do we deploy 4IR technology to ensure that we reach zero fatalities, as a matter of course, within the next five years?

To attempt to answer these questions, we need to understand how the 4IR is generally implemented in the mining industry. The 4IR is achieved through processes, systems and infrastructure. In this regard, mining needs to deploy processes and systems to monitor the movement of people, machines, rocks and slopes underground. The technology of the first industrial revolution gave us tools to allow us to analyse the dynamics of rocks – a case in point being the work of the Russian engineer Stephen Timoshenko. We can thus integrate this technology with AI in order to predict the movement of rocks and ground so as to protect people as well as increase production. This can also be achieved by using cameras underground that are augmented with AI.

Energy consumption should be reduced by deploying energy-harvesting technology underground. Vehicles that are used underground need be autonomous, intelligent and – through the use of IoT technology – connected to other vehicles in order to minimise accidents and maximise production. We need to invest in understanding the psychology of people working underground and their interactions with machines and technology within the context of such a hostile environment.

One technology that has revolutionised the financial sector is blockchain, which could also be useful in the mining industry. Blockchain is used to secure cryptocurrency transactions. It is essentially a digital record of transactions, each one known as a 'block', that are linked in a chain so that one block validates the next. Once a transaction is recorded, its data cannot be altered without changing all the other blocks. This means that the information is always accessible and verifiable.

In mining, blockchain can be used to track the movement of minerals, such as gold and diamonds, and thereby reliably prevent, for instance, the proliferation of blood diamonds – in other words, diamonds that are mined using people coerced or forced into slavery. This would involve digitally stamping transactions relating to gold, diamond, platinum or any other relevant metals or minerals to reliably exclude the use of slave or child labour in mining.

Human irrelevance

It is clear then that we need to understand the overall impact of automation in the mines and its interaction with technology, and particularly so when it comes to its relevance to people. The first industrial revolution created trade unions and united the working class to fight against exploitation. Yuval Harari, in his book *21 Lessons for the 21st Century*, notes that the 4IR is making the working class irrelevant in production and that it would have been easier to organise against worker exploitation in the first industrial revolution than it is against worker irrelevance in the 4IR.[2] This will increase the social burden, which will have to be borne by society and government. A special tax regime, specifically for extractive industries, will have to be developed. The vital question is whether the corporate taxes and royalties mining companies pay to the government are enough. We need to invest in our educational institutions to develop technologies suitable for underground mining, while we also need to understand the psychology of people working with automated machines of the fourth industrial age.

CHAPTER 7

Electricity

In 2019, I received a message from my colleague, David Monyae, who offered a critique that I have been too vocal about the 4IR while South Africa is still fraught with the problems of the second industrial revolution. David made these remarks because the country is struggling with the security of electricity supply, a technology of the second industrial revolution. David's comment came within the context of a political cartoon that showed President Cyril Ramaphosa preaching about the 4IR while South Africa was in the dark because of the failures of Eskom. There is, however, a misconception that for one to understand the technologies of the 4IR, one needs to be able to master technologies from earlier revolutions. But to make a computer – which requires skills stemming from the third industrial revolution – one does not, for example, need to know how to design a steam engine, a technology of the first industrial revolution. The two are entirely independent. One insight that is essential to master is the relationship between electricity and economic growth. Studies in South Africa show that electricity utilisation is correlated with economic growth. It is therefore imperative that we understand electricity supply, because if you compromise the electricity supply, you compromise economic growth.

South Africa's challenge

For more than a decade, South Africa has faced a severe energy challenge personified by intermittent electricity supply. This has often been an argument used by 4IR detractors. How can we possibly focus on the 4IR when we are still struggling with the second industrial revolution? The first bout of rolling blackouts began towards the end of 2007, as Eskom's supply fell behind demand. To prevent this from destabilising the national grid, load-shedding – in other words, taking power-generating units offline for maintenance, repairs or refuelling – was introduced. This did not come as a surprise to many. A crisis of this magnitude had already been predicted in the 1990s when many anticipated that Eskom would eventually run out of power reserves. There were calls at the time – including from within Eskom itself – for a significant investment into the power utility's generation capacity as demand for electricity rose. Then President Thabo Mbeki said at the time, 'When Eskom said to the government: "We think we must invest more in terms of electricity generation", we said no, but all you will be doing is just to build excess capacity. We said not now, later. We were wrong. Eskom was right. We were wrong.'

The power crisis from 2007 to 2008 had a lasting legacy, costing the economy billions of rand. The impact on the economy was devastating, with many businesses left in limbo as mines were temporarily shut down, factories reduced output, and retail outlets were left in the dark. At the time, the National Energy Regulator of South Africa (NERSA) found that measures to provide for a proposed increased electrification programme were lacking but that the growth in the economy had also been too slow.

This, of course, was just the beginning. Since then, we have experienced load-shedding again and again as Eskom struggles to supply electricity amid a confluence of other headwinds. This has ranged from the depletion of dry coal stockpiles at some plants and the tripping of generating units in 2014 to unlawful strikes over wages in 2018. In fact, the power-supply crisis has been acknowledged by credit rating agencies, the National Treasury, the South African Reserve Bank and by many in the business community as the single most significant risk to the South African economy. This is unsurprising when you analyse how things have gone wrong.

In 2015, National Treasury estimated that electricity shortages as a result of electricity supply constraints would see a loss of between 0.5% and 1% from annual real GDP growth. In 2019, South Africa saw its worst bout of power cuts. The GDP figures for 2019 suggest that South Africa fell into a recession in the second half of the year. The CSIR has estimated that the cumulative cost of load-shedding to the economy in 2019 was between R59 billion and R118 billion.

The impact of load-shedding has been easy to quantify, but identifying the cause has been far more complex. The problem with our electricity supply is as a result of a confluence of factors. The evolution of the problem lies in the lack of adequate planning or even a lack of basic understanding of the energy sector, which makes it challenging to manage, let alone maintain. Then, of course, technical issues have been compounded by a fraught political space and allegations of state capture that have permeated throughout the entire country.

Electricity generation

How is electricity generated and why are we struggling to supply it? There are two main ways of generating electricity. The first is to harvest energy directly from the sun using devices such as solar panels, and to convert this energy into electricity. The second, and dominant way, is through a turbine, which involves placing a conductor, such as copper wire, close to a magnet and moving it using either liquid or gas. Eskom generates electricity by moving a giant electric conductor located close to a huge magnet. The elegant theory that explains this magic of moving a conductor and thus generating electricity is called electromagnetism, developed by James Clerk Maxwell. Electromagnetism was the seed for Einstein's theory of relativity, which states that the laws of physics are the same everywhere, and that space and time are united in a single concept called space-time. Simply put, events that occur at the same time for one observer could occur at different times for another. But, if generating electricity is as simple as moving a giant conductor located close to a large magnet, how come Eskom is struggling to achieve this task? To understand this problem, we first need to know how such giant conductors are moved.

There are many ways in which conductors are moved, one being to use coal to heat water, which generates steam that pushes this conductor. This is essentially what a coal power station does. Another way is to use fusion or fission, which are nuclear reactions that release energy. In fission, atoms of radioactive materials, such as uranium, are separated or split, thereby releasing heat, which boils the water, which becomes steam, which then moves the conductor. Fusion is the converse. Hydrogen isotopes (chemical elements within an atom) are fused to form helium, which releases enormous amounts of energy. These processes of fission and fusion are what are used in nuclear power stations.

Understanding the advantages and disadvantages of nuclear versus coal power stations is essential but is not the subject of this chapter. Many believe that nuclear power stations are more environmentally friendly than coal power stations, simply because nuclear reactions do not have carbon dioxide as a by-product, unlike the process of generating electricity in coal power stations. The disadvantage, however, is that nuclear waste is very hard to get rid of and can, in many instances, last for thousands of years.

Once electricity is generated, it is transmitted from the power station to another location, such as a city. At the designated location, it is distributed to houses. In South Africa, and to date, it is Eskom that exclusively generates, transmits and distributes electricity. The task of producing and transmitting electricity is so complicated that municipalities in South Africa have largely abandoned attempts to do so themselves. What remains of their efforts are unused electricity generation plants like the one in Soweto overlooking the UJ campus. It is now used for bungee jumping. The one in Cape Town was demolished in August 2010. However, in 2020, Ekurhuleni and Cape Town were poised to begin generating their own electricity.

Transmission

After the power is generated, it has to be transmitted to a city or town. This transmission is a bulk transfer of electrical power from place to place and is typically done at high voltage in order to minimise technical power loss – in other words, the amount of electricity lost due to the nature of the infrastructure used. For electricity to be transformed from low-voltage generation to high voltage for transmission, a step-up transformer – a device that transfers electrical energy from one electrical circuit to another circuit – is used. Transformers are used to change voltages and currents in transmission lines.

When electricity was discovered, there was much debate as to whether it should be transmitted and distributed via direct or alternating current. Thomas Edison advocated for direct current (DC) to be used, while Nikola Tesla argued for alternating current (AC). DC is when electricity is transmitted at the same current, while AC is when electricity is transmitted at a changing current. To illustrate this, let's consider a hosepipe transmitting water. The current will be the speed at which the water travels, while the voltage is the difference between the pressure at the point the water enters the hosepipe and the pressure at the point at which the water leaves the pipe. If there is no difference in pressure between the entry point and the outlet point, then the water will not move. Ultimately, Tesla won the debate and electricity is now transmitted and distributed using AC.

The transmission network used to distribute electricity is usually comprised of copper-wire powerlines that are supported by pylons, which are the tall, tower-like structures we see across South Africa that are used for carrying electricity cables high above the ground.

How can AI be used to enhance our electricity transmission in the era of the 4IR? Firstly, electricity pylons can be designed using AI in order to maximise their strength while making them cheap to build. Secondly, in an era in which the transmission lines are in danger of being stolen, AI can be used in monitoring these transmission lines to protect them from theft.

Distribution

Once electricity is transmitted and arrives in a city at high voltage, a step-down transformer is used to change it into low voltage before it is delivered from the substation to consumers. In South Africa, power is either distributed by Eskom or by a local municipal company. In some municipalities, local companies – such as City Power in Johannesburg – distribute electricity. Throughout the value chain of electricity supply, the opportunities of using 4IR technologies are extensive. For example, one can use AI to monitor the structural integrity of transformers or the condition of the entire electricity supply grid. Given this, how do we secure the supply of electricity? Naturally, we should secure generation, transmission and distribution.

Security of electricity supply

So, again, why is Eskom struggling to deliver electricity? The first reason the power utility is failing is due to the commercialisation dilemma. This entails the inability of Eskom to collect its revenue. It was reported that in 2019, Eskom was owed more than R28 billion by municipalities. The bulk of this is made up of the R21 billion exclusively owed by Soweto residents. What exacerbates this further is not only Eskom's inability to optimally source coal, but also Eskom's massive debt – R450 billion owed in 2019. Because the bulk of Eskom's revenue is currently used to service its debt, the power utility compromises on aspects of electricity supply that do not have immediate consequences, such as the maintenance of the infrastructure as well as infrastructure development.

The other commercial problem is that the business model adopted after the construction of Medupi and Kusile power stations was based on an expectation that Eskom would be able to charge high tariffs, but NERSA has since put a stop to this. What Eskom requires is a sophisticated treasury department that is skilled in the pricing of options and derivative-investment instruments, enabling it to handle future costs of debt, coal and accounts payable. Furthermore, Eskom requires excellent technical skills at all levels of its operations. It does not help to have the now-former chair of Eskom, Jabu Mabuza, for instance, unable to explain to the Zondo Commission how electricity is generated or what a conveyor belt is.

Super-critical power stations

In 2012, when I was chair of the Education Committee of the Engineering Council of South Africa (ECSA), I learned that the individuals who designed the Medupi and Kusile power stations were not registered as professional engineers in South Africa. To legally practise engineering in South Africa and have the authority to sign off on designs, it is required that the engineer should be registered with ECSA. Even though the engineers who designed Medupi and Kusile might have been brilliant, they were not practising in South Africa, and had worked with locally registered engineers to evaluate the designs.

Super-critical power stations are more modern types of coal-fired power stations that differ from traditional coal power plants in that they use water that is super-critical, in a state that is neither liquid nor gas, meaning that it requires less energy than more traditional methods to change it into steam. The Medupi and Kusile power stations in South Africa are the largest super-critical power plants in the world; nothing has ever been built at the scale of these power plants anywhere else. Super-critical power stations have to operate at specific temperature and pressure, and they have complicated layouts that are difficult to construct.

Why South Africa pursued this complicated and untested technology at this scale is a question that needs to be understood. Additionally, the construction process followed was not based on the turnkey principle – where the project is constructed and handed over to the customer as a completed product – but rather on the order-to-build principle. Why did South Africa not have adequate professional engineers, engineering technologists, project managers and financial specialists to negotiate a good construction deal, especially given the fact that it was a first of its kind? The result is that the cost of the construction of these two power stations, which was expected to have been approximately R200 billion, ended up escalating to R450 billion. This then meant that a technical dilemma became a financial one.

Maintenance

Among other complexities, technical and financial dilemmas naturally lead to maintenance problems. If a company is running short of money and the skills balance sheet is also weak, then maintenance is compromised. Three maintenance strategies are deployed in industry: run-to-failure, scheduled maintenance and predictive maintenance. Run-to-failure is, quite simply, where assets are used until they fail and simply cannot be used any more. The disadvantage of this strategy is that it leads to unplanned power cuts. This strategy is often deliberate in that it saves on costs, but it does not take into account lost production, customer unhappiness, the work required to replace machinery or equipment, and other indirect costs. In scheduled maintenance, assets are maintained and only replaced when they finally reach the end of their life cycle. In this strategy, assets that still work are replaced to prevent unexpected power cuts. When there is a scheduled time for power cuts, it is because the power supplier is performing scheduled maintenance. The maintenance strategy of the 4IR is predictive maintenance. This extends the life of an asset beyond its life cycle.

In contrast to scheduled maintenance, transformers are directly monitored during regular operation in order to anticipate failure. This creates structures that are efficient from financial, technological and operational perspectives. In technical terms, this is akin to neural networks, which we discussed in Chapter 1, and which form part of deep learning, a central tenet of the 4IR.

Neural networks use a set of algorithms that mimic how the human brain works, particularly in terms of recognising patterns. Here, Eskom would be able to estimate when its transformers will fail and replace them beforehand.

If you can predict when this will happen, you can get optimal use out of them. The questions that we ought to ask are: what strategy is Eskom using to maintain its assets, and what are the financial implications?

Simplifying complexity

In 2019, President Ramaphosa announced in his SONA that he would be breaking Eskom up into three new state-owned entities (SOEs): Generation, Transmission and Distribution. This idea is not new. In the previous decade, the South African government established Electricity Distribution Industry Holdings (EDI), which was intended to consolidate all electricity distribution under one umbrella. Ultimately, this notion was abandoned. This time, we need to pursue a solution that is optimal for South Africa and create structures that are efficient from financial, technological and operational perspectives as well as improve the security of electricity supply.

Running a power-utility company is a complicated exercise with multiple objectives. At the moment, the type of leadership we need is one that can optimise and balance multiple objectives that are sometimes in conflict with one another. Only focusing on the economic objectives of Eskom acts against intensifying maintenance and technical capacity. Vilfredo Pareto, a French-born engineer of Italian heritage, studied the concept of managing multiple objectives in depth. He observed that the ability to choose the trade-offs between all these competing objectives or goals is subjective and, therefore, requires one to be pragmatic. Given all of this, what is to be done? Firstly, South Africa needs to be decisive and understand that in order to sort Eskom out, it needs to invest, and to do this will require financial sacrifice. Secondly, South Africa needs to capacitate Eskom – from the board to the lowest levels – from a technical standpoint. Thirdly, South Africa needs to adopt a maintenance strategy and stick with it.

PART 2

Data

This section focuses on data. Data in itself has various meanings, ranging from the data that helps us connect to the internet, to the data that companies and governments collect about people, such as their demographic information, for example. To define the concept technically, data is various distinct pieces of information. However, in this age, when data has become currency, certain safeguards need to be taken. As we begin to share more about ourselves on social media platforms, such as Facebook or Instagram, we need be aware of what the cost may be to us and where our data ends up. As we delve deeper, we must also begin to realise the importance of cybersecurity. By the same token, it is also imperative for us to build databases to preserve our history. This section looks at many of the questions and issues surrounding data.

CHAPTER 8

Data privacy

It may seem absurd, but we are all far too quick to give away our personal data. By giving in to clickbait to find out what I might look like in 40 years, I am effectively giving an external website access to all my personal data on Facebook. Interestingly, the caveat that they will not post to my Timeline without my permission is often the only reassurance we are looking for. After a few clicks, it turns out that this site thinks I might look like Clint Eastwood, as far-fetched as that may seem. Yet, for a moment of amusement and possibly a couple of Likes on my Timeline, I allow this site access to everything on my Facebook account with little knowledge of who is behind the site or for what it might use my information.

A trade-off for our data has thus become commonplace. You would be hard-pressed to find a news site that doesn't ask you to accept cookies – which store bits of information about your interactions with the website – or a site that doesn't ask you to connect with or to log in to it through your Facebook account. This trade-off permeates just about every aspect of our online communications. As British mathematician Clive Humby said, 'Data is the new oil.' In 2019, one of South Africa's biggest banks, First National Bank (FNB), partnered with the Entertainer app, which provides discounts or deals on restaurants, travel, health and beauty. No doubt a perk for customers, the app also offered FNB invaluable insight into users' spending habits. Medical-aid giant Discovery Health works somewhat similarly through their Vitality Rewards programme, which compensates users for completing health checks, doing an HIV test or attending the gym regularly. Like FNB, Discovery stores this data and uses it to keep track of its users. This idea of storing data for rewards has been adopted by the Chinese government, which uses the Social Credit System to control and manage people through machines. Following the outbreak of the coronavirus at the beginning of 2020, this proved a useful tool to track the spread of the virus and provide adequate treatment. By using voice- and facial-recognition technology in conjunction with big data, the Social Credit System can identify people, pinpoint their location and learn about their prior movements, as well as keep track of their health. The system can then categorise this information and produce daily reports.

These, of course, are instances where the trade-off seems worth it. People are happy to relinquish their personal data if it can keep them healthy and contain a global pandemic. They may even be happy to relinquish their data by using an app in exchange for a free meal. The question, however, becomes: what constitutes an adequate trade-off? There is an urban legend that between the 16th and 18th centuries, African people exchanged their land for mirrors. This legend was aptly dramatised by Jomo Kenyatta, the first president of Kenya, when he said: 'When the missionaries arrived, the Africans had the land and the missionaries had the Bible. They taught us how to pray with our eyes closed. When we opened them, they had the land, and we had the Bible.' By all definitions, an exchange of land for a mirror or any book is not a good business deal. But this may, of course, depend on the significance placed on these objects – particularly at the time. It is hard to reconcile this notion today, but, at the time, mirrors were seen as luxury goods; they represented opulence and an ideal. We know now that this was an irrational trade. For anyone to participate in any form of trade, one needs to understand the concept of the value and the utility of goods.

Utility theory

'Utility' is an economic concept that means the usefulness of particular goods or services. In this way, utility is a measure of the value of specific goods and services. In terms of utility theory, mirrors have significantly lower utility than land. Rational beings can only accept the exchange of products and services if the transaction results in additional satisfaction, which in economics is known as 'marginal utility'.

For example, if Mbali has R5 000 and Thendo has an iPhone, then Thendo will accept the R5 000 in exchange for the iPhone if he thinks the relative value of R5 000 is higher than that of the iPhone. Conversely, Mbali will accept the iPhone in exchange for the R5 000 if she thinks the relative importance of the iPhone is higher than that of the R5 000. So, for this transaction to happen, there has to be a difference in the perception of value, a concept referred to as the 'asymmetry of the understanding of value'.

The difficulty of using utility theory to understand the relative value of different products – the iPhone versus cash in the example above – lies in comparing the two products. The easiest way of distinguishing them is to convert them into their monetary values and then compare them in financial terms. Converting goods and services into monetary benefits brings to the fore a subjective concept that changes over time called 'market value'. A new iPhone today costs more (that is, it is more valuable) than the same iPhone will cost in two years' time. The market value makes the idea of relative value subjective. Ultimately, however, utility theory is based on the decision-maker's preference for either the R5 000 or the iPhone.

Unfair transactions

Sadly, however, we live in an era in which many in our society are making transactions similar to the exchange of mirrors for land; an age wherein we are told that many vital goods are demonetised – in other words, free of charge. Of course, there is no such thing as a free lunch – and there never was! We are told that Google gives us the Gmail account for free. We are told that Facebook gives us WhatsApp, Instagram and Facebook accounts for free. We are told that we no longer need to pay for video conferencing because we can make video calls for free using WhatsApp. Of course, for us to make such video or voice calls using WhatsApp, we need data – which we can buy from Telkom, MTN or any South African telecoms company – so this initial expense already means that there is a charge involved.

However, demonetised services such as Gmail, WhatsApp and Facebook accounts come with an additional cost, which ensures that they are not free, because they are given to us in exchange for our personal data – information about us. Once we subscribe to these accounts, we surrender our data and, consequently, we are monitored 24 hours a day, not by a person but by sophisticated software – and sophisticated software is more efficient in surveillance than a well-trained person. Consider how much information you share about yourself online. These organisations know our movements, what we like and do not like, where we might be going on holiday, what we eat and what websites we visit, for instance. They then sell this information to the highest bidder without our knowledge. In 2019, around 4 000 documents leaked from Facebook to American news network NBC showed that its CEO, Mark Zuckerberg, controlled competitors by treating user data as a bargaining chip. According to an in-depth news report, the company allegedly granted access to partner companies that spent money on Facebook or shared their own data with Facebook, as well as with app developers considered personal friends of Zuckerberg. The impact of this is far-reaching. In 2018, Zuckerberg testified before Congress after it was revealed that British research and consulting firm Cambridge Analytica had used the personal data on around 87 million people's Facebook profiles without their consent for political advertising.

How did this happen? Cambridge Analytica created a survey for academic use in which thousands of Facebook users took part. Facebook, however, allowed the app to collect the personal information of all those in the participants' Facebook social network. With this data, Cambridge Analytica was able to create psychographic profiles for political campaigns, which included the type of advertisement that would be most useful to users. As Zuckerberg told Congress, 'Facebook is an idealistic and optimistic company, but it is clear now that we did not do enough to prevent these tools from being used for harm as well. That goes for fake news, foreign interference in elections and hate speech, as well as developers and data privacy. We did not take a broad enough view of our responsibility, and that was a big mistake.'

If you are walking in a shopping mall and you realise that someone is following you and is recording all your movements, you would be terrified. Why then are we not terrified about being tracked by Google or Twitter or Facebook's artificially intelligent software? We are not terrified of the online follower (software agents) – as opposed to an offline follower (human agents) – because the online follower pretends that it is not there. We neither hear its footsteps nor see its shadow reflected on a windowpane, because footsteps and shadows are only digitally as opposed to physically present. Human beings did not evolve to perceive digital; they evolved only to recognise physical steps and shadows. So it is that 'borderless' companies extract much more value from those who subscribe to them than the amount of value they give back to users. This means that those of us who are subscribed to these technologies are irrational, even though we might not know it yet, because we benefit far less than what these companies gain.

Why does unfair trade of data exist?

For trade to exist, there has to be a different or asymmetrical perception of value between the two parties involved about the items being exchanged. In our example above, we saw that Mbali needed an iPhone more than Thendo did, and Thendo needed Mbali's R5 000 more than the iPhone. This is a necessary condition for a deal to happen, but it may not necessarily be a sufficient condition. Why do people exchange personal information for a social-network account? Firstly, because they are not able to accurately calculate the relative utility between personal data and having social-network accounts.

Nobel Prize winner Daniel Kahneman describes this practice of not using calculations in order to make decisions in his book, *Thinking, Fast and Slow.*[1] Kahneman claims that human beings never make calculations in order to make decisions but, instead, use shortcuts known as heuristics, and this results in biased conclusions. Another reason why people make such an exchange is that they see the benefits of being connected to social networks as exceeding the disadvantages. This is because social networks activate the addictive parts of our beings, targeting our needs to be socially connected, allowing us to pry on others, and cultivating a false sense of fame, for instance. As Adam Alter, author of *Irresistible: The Rise of Addictive Technology and the Business of Keeping Us Hooked*, said: 'When someone likes an Instagram post or any content that you share, it is a little bit like taking a drug. As far as your brain is concerned, it is a very similar experience. Now the reason why is because it is not guaranteed that you are going to get likes on your posts. And it is the unpredictability of that process that makes it so addictive. If you knew that every time you posted something, you would get 100 likes, it would become boring really fast.'[2] This is unsurprising when you consider that US-based marketing agency Mediakix found that people spend more time on social media each day than they do on eating, drinking and socialising combined.

A study by another marketing agency, Smart Insights, found that people publish 3.3 million Facebook posts, 448 800 tweets, 65 972 Instagram photos and 500 hours of YouTube video in just one minute. This has been described as the attention economy, which is essentially the supply and demand of people's attention – the commodity used on the internet. Social media makes such good use of social feedback in the form of a Comment or a Like to induce a short burst of euphoria that we wait in anticipation for the next one. As the Facebook Like button co-creator Justin Rosenstein told *Vice* a few years ago, 'These are our lives. Our precious, finite, mortal lives. And if we're not vigilant, computers and mobile devices will guide our attention poorly.'[3]

Security

What are the implications for Africa's security if so many people who have subscribed to these technologies are being monitored? Well, many of Africa's top leaders in politics, business and society have indeed subscribed. When these individuals send messages to one another through Gmail accounts, those messages often travel from South Africa to the US and then back to South Africa. This means that such messages can be intercepted and used in a manner not in the best interests of the country.

So, what have other countries done to prevent such breaches of security? China, for one, has barred these companies from operating within their borders. Gulf countries, such as the UAE, have limited certain services like WhatsApp calling. Even though the authoritarian and economic-protectionist explanations of this ban cannot be discounted, the security aspect is more persuasive. For example, neither Gmail nor Facebook nor Twitter nor WhatsApp accounts work in China. Anyone who has travelled to that country will have experienced the extreme withdrawals of not having access to these accounts. This means that not only are we selling our data to these companies, but that we are also addicted to selling our data to these companies.

An interesting historical parallel can be drawn here in the sense that China, having fought in the Opium Wars against Great Britain (who had imposed the opium trade on China), in the mid-19th century, experienced first-hand the effects of huge swathes of its population becoming addicted to a foreign-introduced fixation, to the detriment of its development. As a result of the wars, China lost its position as the largest economy in the world, and it also lost Hong Kong to Britain for 150 years.

More than a century and a half later, China was again faced with a similar threat in its people becoming addicted to the foreign-developed and highly addictive stimulant of social media. Again, it stood to lose out, not only in the negative effects social media might have on the populace, but also in terms of the fact that the country would not be in control of these platforms as well as of people's data. This would have resulted in its people's movements and data being monitored by companies based in other countries, to the detriment of China's national security. What China did in response was to create its own versions of these technologies: Baidu instead of Google, WeChat instead of WhatsApp and Weibo instead of Twitter.

The Chinese response to social media is not, however, replicable in Africa for several reasons. Firstly, unlike in China, where there was a demand for social media platforms in a language other than English, in Africa, people are likely to use English-based social media and, as a result, it would be more difficult for new social media platforms developed in English in Africa to compete with established ones. Secondly, should African countries choose to develop their own social media platforms in African languages, the sheer number of diverse languages across the continent (about 1 500 to 2 000 different languages), would make this impractical, if not impossible. Thirdly, the skills dilemma prevents Africa from developing its own social media platforms.

However, these challenges are not insurmountable. For example, if African countries were to ban certain international social media platforms from operating within their borders, like China did, our own home-grown solutions would stand a better chance of being adopted. Furthermore, AI could be used to tackle some of the challenges of translation.

Control systems

As we saw in Chapter 5, control systems are essentially devices that are used to control automated systems, comprising of a mechanism to collect data, a model to understand that data, and a feedback mechanism to ensure the relevance and accuracy of the system's decisions. For example, in aircraft, control systems are used to identify the optimum altitude at which there will be minimal turbulence. Control systems are also widely used in factories, from breweries to car-manufacturing plants.

The first task in controlling any system is to monitor it. So it is that social-networking companies track the behaviour and movements of people, and – if we take the idea of control systems to its logical conclusion – thus control and influence people.

This notion of controlling people is known as nudging. Richard Thaler was awarded a Nobel Prize for his theory of how people are nudged to behave in a certain way; a 'nudge' meaning something that alters people's choices and behaviour in a predictable way without forbidding any options or significantly changing their economic incentives.[4] Yet, in the context of our data, we must ask whether we are happy with some company controlling our lives here in southern Africa remotely from California. Is this what South African leaders, such as Steve Biko and Robert Sobukwe, died for? Is this what Govan Mbeki and Walter Sisulu went to jail for? Is this what Chris Hani and Oliver Tambo went into exile for?

What is to be done?

What then is to be done to ensure that the large-scale surrender of personal data to companies domiciled in the US, or any other part of the world, does not result in a new kind of colonisation of this particular type? Firstly, Africa needs to create a regulatory framework around the use of private data across our borders. In some regard, Africa is already in the process of implementing regulations around data privacy. The Protection of Personal Information Act, which is expected to come into effect in South Africa before 2021 but which has fallen behind schedule, is aimed at protecting people from harm – the theft of either their money or identity. The Act provides for penalties of up to R10 million and even jail sentences of up to ten years, subject to the gravity of the offence. Secondly, Africa needs to educate its population on the need to understand the concept of exchanging personal data for an email or social network account and the implications thereof. This should include educating South Africans at all levels, from primary to secondary, tertiary and even adult-basic education. Thirdly, and perhaps more challenging, Africa needs to start thinking about creating its own national and even continental social networking companies. In this regard, Africa must study the managed services providers (MSPs) and learn how to price them in order to achieve digital sovereignty. The experience of China and, to some extent, Russia, will be useful in this regard. Fourthly, Africa needs to get its political leaders to stop using 'free' email accounts provided by popular, large international companies, because this compromises the continent's security. The measure of how serious a country is when it comes to data security is to gauge how many of its top leaders use Gmail accounts as opposed to sovereign accounts – official email systems operated and controlled by government, which ensure the country's ownership and protection of data – in order to conduct their business. Much the same is true in the field of education: the measure of how serious a university is can be determined by how many of its leaders use Gmail accounts rather than university accounts.

Africa has to fight for its sovereignty. Whoever controls data controls the future, and Africa must invest in data technology. And to protect its data and sovereignty, the continent needs a transactional system that is rational, reciprocal and fair.

CHAPTER 9

Digital heritage

Three months after a fire engulfed the Notre-Dame Cathedral in Paris, the website Europeana launched an online exhibition, 'Heritage at Risk', detailing the use of digital technology in assisting the cultural-heritage sector in fulfilling the restoration challenges in the 21st century. The fire, which broke out in April 2019 just beneath the roof of the cathedral, destroyed the building's spire and most of its roof, the smoke damaging relics and artworks, and utterly destroying parts of the exterior art. More than six months later, the iconic building was still at risk of collapse, despite efforts to restore it. As Mariya Gabriel, European Commissioner for Digital Economy and Society, put it: 'The fire of Notre-Dame highlighted once more the need to preserve, record and protect our European cultural heritage. Reactions all over Europe show how important it is for our citizens and society. More than ever, we are making progress in harnessing the power of digital for our cultural heritage, but we must keep working together to support the sector in this endeavour.'[1] Social media was flooded with photos of the damage done to the cathedral, and philanthropists vowed to donate hundreds of millions of euro to the restoration. French President Emmanuel Macron treated the fire as a national emergency, saying, 'Like all of our fellow citizens, I am sad tonight to see this part of us burn.' The digitisation of cultural heritage has proven to be important in the restoration of this monument. A few days before the fire at Notre-Dame, the European Commission's declaration of cooperation on the digitisation of cultural heritage saw 25 European countries pledging joint initiatives for the 3D digitisation of cultural-heritage artefacts, monuments and sites.[2]

Histories, biographies and the vantage point from which these stories are told are central to our understanding of events and making sense of them. A significant dimension of the decolonisation debates that occurred around 2015 with the #RhodesMustFall movement centred on the need for Africa to reclaim its right to tell its own stories, to explain its history, and to take its place alongside other dominant knowledge systems. This debate is intrinsically related to the internet. Over the last two decades, the internet has become the first port of call for a search on any topic. One might wonder what this has to do with digital heritage. In the past, for example, Western narratives shaped much of what surfaced in a search on the internet. A case in point has been the recent reopening of the Dr Neil Aggett case. The careful documenting online of the atrocities of the past, including the availability of the TRC's papers, enables the story to be told and for these tragedies of apartheid to become a discourse that is shared worldwide. The point is that digital access to materials is the responsibility of all – the alternative is to have the viewpoints of outsiders as central, which then has the potential to make the significant a mere footnote in history. In South Africa, there is, for example, South African History Online, which is a repository of South African and African history that aims to give voice to previously marginalised voices in order to interrogate the skewed way in which history, cultural heritage and the anti-apartheid struggle, in particular, has traditionally been depicted. It provides us with an authentic perspective. The South African History Archive (SAHA) is an autonomous library devoted to recording, supporting and promoting greater consciousness of past and current struggles for justice. This initiative is funded privately rather than by the government, yet it is a repository of critical information that is accessible at the press of a button. Similar initiatives include the Nelson Mandela Centre of Memory, which is working on digital archives. The O'Malley Archive also offers a rich understanding of the history of South Africa, particularly in terms of the anti-apartheid struggle and from the viewpoint of activists who played a central role in it. The value of these archives is unquantifiable, yet we have seen archives not being preserved despite the technology available.

An important question to ask is: how does Google source information, and should it be viewed as a reliable source? If we did not have initiatives such as the digitisation of our history from the perspective of the other, a simple search for many of our South African leaders would show them up as 'terrorists' or 'persons of interest' for the previous government pre-1994. Critical to reclaiming the telling of African stories is to ensure that they are actually told by us rather than us relying on Western perspectives. As Patrice Lumumba said, 'Africa will write its own history.'

This, of course, is one instance of the importance of heritage data. Our history does not just need preservation; it also needs telling in a way and through a medium that provides access to all. When trying to understand the 4IR, it is imperative to trace the history that led to this moment. If we did not have three previous industrial revolutions, we could not have a fourth one. We often downplay the importance of history, yet without history, how do we know where we have come from and/or where we are heading? In the 4IR, this is no longer akin to stories told around a fire, passed down from grandparents to grandchildren. Nor is it simply what is found in history textbooks. Digital heritage is the use of digital media to preserve cultural or natural heritage, which leaves room for interaction and shared knowledge. As American author Michael Crichton said, 'If you don't know history, then you don't know anything. You are a leaf that doesn't know it is part of a tree.'

Open-source

Twenty-two years ago, when I was a doctoral student in AI at the University of Cambridge, I had to create all the AI algorithms I needed in order to understand complex phenomena related to this field. As we have already seen, AI is computer software that performs intelligent tasks usually done by human beings, while an algorithm is a set of rules that instruct a computer to execute specific tasks. At the time, the ability to create AI algorithms was more important than the ability to acquire and use data. I had to know how to create these AI algorithms before I could use them. The world today is radically different, of course, and now you can type in a query in Google and prompts appear to almost shape or predict your search query.

The landscape has thus changed. Google has created an open-source library called TensorFlow, which contains all the developed AI algorithms. Open-source refers to a program with source code that is made available for use or modification by other users or developers. TensorFlow bundles deep-learning models and algorithms to build apps. It has been used to prevent blindness by helping doctors screen patients for diabetic retinopathy. It has also been used to search for new planets. While Google wants people to develop apps using their software, the pay-off for Google is that it will collect data on any individual using the apps developed by TensorFlow. If you do a Google search for a holiday in the Mediterranean, for example, you will suddenly see advertisements related to such a vacation appearing on

Facebook and other social media platforms. Today, an AI algorithm is not a competitive advantage, but data is. The WEF calls data 'the new oxygen', while British mathematician Clive Humby calls it 'the new 'oil'. This is not any different to what banks or medical aids do: they use data gathered from clients or patients to predict consumer spending patterns and to build that into the development of new products or predictive models for health risks.

Population growth

The population of the African continent is increasing faster than in any region in the world. Africa has a population of 1.3 billion people and a total nominal GDP of US$2.3 trillion. In fact, according to the Institute of Security Studies, Africa's population is expected to increase by roughly 50% over the next eighteen years and will account for nearly half of global population growth over the next two decades. In effect, this increase in the population will result in an increase in data. If data is the new oil, this growth is akin to the rise in oil reserves. Even oil-rich countries, such as Saudi Arabia, will not experience an increase in their oil reserves. So, how do we as Africans take advantage of this enormous amount of data? There are two categories of data: heritage and personal data. Heritage data resides in society, whereas personal data resides in individuals. Heritage data includes data gathered from our languages, emotions and accents. Personal data includes health, facial and fingerprint data.

Data companies

Facebook, Amazon, Apple, Netflix and Google are data companies. They trade data to advertisers, banks and political parties, for instance. For example, the controversial consulting firm, Cambridge Analytica, harvested Facebook data to influence the presidential election that potentially contributed to Donald Trump's victory in the US elections. Google collects language data to build a service called Google Translate, which translates from one language to another. This application claims to cover African languages, including isiZulu, Yoruba and Swahili, but it is clear that Google Translate is not effective in handling African languages compared to European and Asian languages. Now, how do we capitalise on our language heritage to create economic value? We need to build our language database and create our own versions of Google Translate. Currently, the translations provided by Google are so off base that they make no logical sense to a reader, so there is thus a need to train Google Translate in grammar, local inflexions and the multiple meanings of words in order to avoid a machine translation that recognises only the literal meanings of words. The question is whether Google Translate has any depth of understanding and whether that depth should come from primary users of the language.

African database

An important area of Africa's digital heritage is the creation of an African emotion database. Different cultures exhibit emotions differently. Yet, reading emotions accurately is essential when it comes to areas of technology such as the safety of cars and aircraft. If we can build a system that can read pilots' emotions, we will be able to establish whether a pilot is in a good state of mind to operate an aircraft, and this, in turn, can improve the safety of passengers as well as prevent damage to aircraft. To capitalise on the African emotion database, we need therefore to create a databank that will capture emotions of African people in various parts of the continent and then use this database to create AI apps that read people's emotions. For example, Mercedes-Benz has already implemented a feature called 'Attention Assist', which alerts drivers to fatigue. This system uses audio to make the driver aware of behaviour that could be influenced by drowsiness, and it is based on an analysis of steering-wheel movements to create a driver profile, thus identifying when movements are out of the norm.

Another crucial area of Africa's digital heritage is the creation of an African health database – AI algorithms can diagnose diseases better than human doctors can. However, these AI algorithms depend on the availability of data. To capitalise on this opportunity, we need to create a program that collects such data and uses it to build algorithms that will augment medical care. After all, the corporate world uses data for pricing by looking at patterns across the value chain. For example, in the development of South Africa's National Health Insurance, the success could be leveraged off the sound collection and utilisation of data to determine patients' needs, histories, the distribution of resources and the allocation of physical sites.

The integrity of the data could enhance the usage of medical algorithms to standardise and customise health care.

One of the latest technological developments is the rollout of intelligent personal assistants – in other words, devices that can take voice instruction. Google has already developed Google Assistant; Amazon has created Alexa; Apple has built Siri; and IBM has developed Watson. These are all very effective but do not handle African accents well. Trevor Noah famously made fun of this in one of his shows when he acted out Siri misinterpreting an Afrikaans accent. My own devices have to be trained to recognise my name. We can, however, enhance these devices by including emotion-detection algorithms, making them more sensitive to different accents, especially abundant and diverse African accents. But, for us to capitalise on the oversight, we need to create a database of African accents and use this to build intelligent personal assistants that can understand African languages.

Facial-recognition algorithms do not work very well for African faces either. Police in the US use IDEMIA, which scans faces, recognises them and identifies the individual by means of algorithms, yet results from the US National Institute of Standards and Technology have indicated that two of IDEMIA's algorithms were pointedly more likely to misidentify black women's faces than white women's, black men's or white men's faces. Where IDEMIA's algorithms erroneously identified the faces of white women only one in 10 000 times, it incorrectly identified black women's faces one in 1 000 times – so, ten times more frequently. One of the reasons for this is the limitations of African facial databases. Another is the suboptimal data collection for African faces, which are different from Asian and European faces. A third reason is that we have not designed AI algorithms for facial recognition from the African perspective, but rather from the European and Asian perspectives. There are often errors, for example, when Facebook suggests people to tag in photographs. Companies such as Facebook are collecting vast amounts of data from African people who have Facebook accounts. Facebook is, however, only in the data-collection stage, and enabling a production model is yet to be implemented. It is clear then that we need to create our own facial database so that the Department of Home Affairs (DHA) can, for example, use this to improve security at points of entry into our country. Currently, for the Smart ID card, the DHA is imaging the front of people's faces only. For a facial database, imaging is required for the sides of the face as well. In a way, Facebook is building this, with our help, as we upload our images. In these images, the emotions of people photographed can also be gleaned, contributing further to another aspect of Facebook's database.

All of this is, of course, much more complex than may be suggested here, simply because Africa is not homogenous. It is estimated that there are at least 3 000 ethnic groups and up to 2 000 languages spoken in Africa. How do we posit solutions that speak to the entire continent? We have already seen that Siri and Google Translate are not as effective in African countries, so there is a fundamental need to create our own platforms. But what does this look like? How would such a platform be able to adapt from Nigeria to Egypt to South Africa? We need to find the solutions ourselves.

Data as heritage

There is much heritage and personal data that we, as Africans, can collect and monetise to derive economic value. Some of these data include images of the iris of the eye and fingerprints, both of which are extremely valuable for building biometric security systems. However, for us to be ready, we need to develop sets of skills in order to collect and analyse these data effectively. This includes data analytics and AI-algorithmic skills. To use open-source AI algorithms, one requires some understanding of programming. The data-analytics skills should go beyond the basic statistics courses that we often find in our universities and must include advanced topics, such as signal processing as well as the capability of handling incomplete and imperfect data sets.

In 2017, Kenyan software programmer Kennedy Kanyi came up with BraceScript, a simple coding language launched in Africa to teach local kids how to code and to develop future software engineers.[3] In South Africa, many schools have introduced coding, but this is often limited to the more affluent schools. It may be even more critical for our children to learn the algorithms underpinning coding. While some initiatives are underway, there is a shortage of tech talent on the continent.

Increasing the capacity to collect data

How does Africa then increase its capacity to collect and analyse data? Firstly, we should introduce national databanks that collect these data, but we should do this in such a way that we protect data security and privacy. One way of achieving this is to expand the mandate of organisations such as Stats SA to include the gathering of personal and heritage data, in addition to gathering and analysing economic data and performing the national census. Regional organisations, such as the SADC and ECOWAS, must create regional databanks that collect and monetise data. At the continental level, the AU should establish databanks that will consolidate and monetise the continental database. New opportunities for this arise with, for instance, the AU's approach to the African Union Passport. On creating such databanks, we should bear in mind, however, that any given database will inevitably be incomplete and imperfect because you cannot know all the possible variables, and measurements are often not completely accurate. We, therefore, need to capacitate data-gathering organisations with the competence to analyse the data at hand. If we are able to explore the vast heritage and personal data of the 1.3 billion African people, then we can create value for Africa through the 4IR.

There are, however, ethical implications in the collection of individual data, and we need to take these into account. The recent Cambridge Analytica scandal illustrates how damaging it can be when there are data breaches. In Great Britain there has been, for example, a discussion on using AI to identify possible criminal behaviour based on behavioural trends. This process, however, has the potential to be abused because there are several underlying assumptions that could or could not be plausible. As an identification tool, AI has been used successfully in many countries, especially with the widespread usage of cameras, sensors and other monitoring devices. Just as with any technology, it needs to be managed carefully and have the appropriate ethical foundations built in before it can be used widely. Thus, the ethical usage of data, AI and algorithms without compromising human beings is essential.

CHAPTER 10

Cybersecurity

In 1953, Nelson Mandela came up with the M-Plan – a decentralised strategy of organising and executing the struggle efficiently in all areas, whether urban or rural. As Oliver Tambo put it, 'We had to forge an alliance of strength based not on colour but on a commitment to the total abolition of apartheid and oppression; we would seek allies, of whatever colour, as long as they were totally agreed on our liberation aims. The African people, by nature of their numbers, their militancy, and the grimness of their oppression, would be the spearhead of the struggle. We had to organise the people, in the town and countryside, as an instrument for struggle.'[1] This was the precursor to the African National Congress (ANC) operating underground. Embedded in the M-Plan was a matter of secure communication. It was important that, in this M-Plan, apartheid agents and its infrastructure did not know what was happening in the community.

This process of concealing information from some, even though it remains visible to others, is known in computer science as 'encryption'. Madiba's M-Plan, therefore, involved encrypting the activities of the ANC. When one sends an SMS from Qunu to Duthuni, that message is generally not encrypted, and, as a result, sophisticated technology can be used to read its contents. The only way to maximise the chance that this message won't be understood by anyone other than the intended recipient is to encrypt it. Messaging-technology companies, such as WhatsApp, claim that their messages are secure and cannot be decrypted by third parties. Similarly, under the M-Plan, messages were transmitted from national to local levels in an 'encrypted' manner, without using people's real names.

Cybersecurity threats

In recent years, there has been an increase in cyberattacks, which are malicious and deliberate attempts to breach an information system. According to a 2019 report by PricewaterhouseCoopers (PwC), cyber threats are the fifth-largest threat to global growth, particularly as companies increasingly embrace digitisation, automation and AI. Malware (malicious-software) attacks in South Africa increased by 22% in the first quarter of 2019, compared to the first quarter of 2018, according to global cybersecurity company Kaspersky Lab. This translates to about 13 842 attempted cyberattacks per day, just under 577 per hour, or more than 9 per second.[2] According to a 2019 Mimecast report, 88% of South African organisations had experienced a phishing attack in the previous twelve months. Eight out of every ten South African organisations had suffered an impersonation attack, with 63% reporting an increase in such attacks; and 42% of local organisations had experienced a ransomware attack in the past twelve months, compared to 23% in the prior twelve months.[3] In October 2019, the City of Johannesburg reported a breach of its network, which shut down its website and all e-services, after receiving a bitcoin ransom note from a group called the Shadow Kill Hackers. Simultaneously, several banks reported internet problems also believed to be related to cyberattacks.[4]

Since most sectors have moved towards the digitisation of information, cybercrime and targeted attacks breaching cybersecurity have become more commonplace, with the potential to negatively impact society. So, while digitisation has created effective and efficient integrated systems, the priority needs to be to protect these systems from attack.

A simple example of low-level cybercrime is the cloning of Facebook profiles, with offenders sometimes utilising the features of the platform to solicit money from friends or post inappropriate content. Given the relationship between smartphones and the seamless integration of multiple apps on these devices, when one loses their smartphone due to theft, for example, and it lands up in the wrong hands, it provides the ideal opportunity for criminals to commit severe breaches in cybersecurity. Despite the inclusion of numerous security features on devices such as laptops, tablets or smartphones, ranging from passwords to fingerprint-identity sensors, the sophistication of cybercriminals is such that they can penetrate these security measures, resulting in significant damages.

As a result, large corporates – or other users with multiple employees – have to ensure the robustness of their IT systems in order to block, for example, emails that seem innocuous but are maliciously sent to users requesting them to click on a link that then downloads malware. Perhaps the most infamous example of malware is the Melissa virus, which infected thousands of computers worldwide towards the end of 1999. It was spread via email with the subject line, 'Important Message From [name of the infected user]', and a Microsoft Word-document attachment that, when opened by the user, would result in the email being mass-sent to the first 50 people on the person's contacts list. It is estimated that this caused losses of US$80 million to companies and individuals.

Another form of cyberattack is phishing, which involves cyber criminals sending emails that purport to be from reputable individuals or companies, with the purpose of getting the user to divulge information such as passwords or credit-card details. Cyberattacks can also come in the form of ransomware, a type of malware that locks users out of their systems or threatens to share their personal information unless they pay a ransom. In recent years, cryptocurrency – an otherwise-legitimate form of digital currency – has emerged as the most-used payment method as demanded by cybercriminals of their victims.

Another way that cyber criminals target companies and other entities is through distributed denial-of-service (DDoS) attacks, which involve intentionally disabling a computer network or website by inundating it, or its host, with data sent simultaneously from various computers, resulting in the network or host crashing or becoming temporarily unusable.

Consumers' personal information is also often traded as a commodity by legitimate companies that have harvested their data. While this may not constitute cybercrime, it does highlight the need for individuals to be cautious about providing private information to others. For instance, opening an account with a department store requires you to provide details of your address, income, banking information, access to credit ratings and other information. Millions of consumers offer this data and then are surprised when they receive offers from other companies, phone calls from telemarketers and SMSes from political parties. There is much money to be made from selling databases. While these may not be conventional cyberattacks, they do illustrate that the existence of personal data on multiple databases can be risky. It is the company's responsibility to ensure this data is secure. In 2016, an under-publicised hack against Uber occurred, through which the data of drivers was accessed. Uber succumbed to the request for a ransom payment simply because the required precautionary measures were not in place. The resulting reputational damage to Uber was high, and the dent to its finances was significant.

Another area in which individuals' personal data may be vulnerable is in health systems. Cyber criminals can obtain people's personal information either through accessing their medical records, stored in databases, or their medical insurance claims. The increasing sophistication of medical technology enables the transmission of data from a patient to a medical professional. For example, glucometers (used to monitor the blood sugar or glucose levels in diabetics) or pacemakers (which monitor heart rhythms) are connected to the internet, thus increasing the risk of a third party being able to access the data. While these kinds of technology are viable and efficient health solutions, they are susceptible to attack.

In the field of logistics management – which involves the coordination of the transportation of goods – delivery schedules and routes are managed through the use of IT systems, which, if breached, can provide criminals with the information they can use to plan their attacks. In South Africa, for example, we have had high incidences of targeted attacks on cash-in-transit vehicles. Despite the fact that the companies that operate these vehicles track their movement using GPS, with dispatchers and drivers in constant contact, there appear to be frailties in the system, given how these robberies are undertaken with such precision.

In 2019, the personal information of cruise passengers, crew and employees were compromised after an unauthorised party gained access to the email accounts of employees working for Princess Cruises and the Holland America Line corporation.

There are, of course, far more harmful consequences of malicious cyber activity, such as those that result from cyber warfare. In 2019, the US, UK and Estonia condemned cyberattacks against Georgia, part of the former Soviet Union, by Russian military intelligence at a closed-door meeting of the UN Security Council.[5] That same year, the US retaliated against Iran's downing of a surveillance drone with an offensive cyber strike to disable the computer systems used to control rocket and missile launches. This sparked a debate about the vulnerability of countries' command-and-control systems.

As users of IT systems, Cloud-based platforms and multiple apps, we all have to ensure that we have a cybersecurity plan in place. This includes taking preventative measures, such as using encryption, antivirus software and firewalls, for example. These measures should be customised according to our individual use of cyber technologies and the associated risks that come with this.

However, having such a plan in place does not provide you with any guarantee that you will not come under attack. Therefore, it is also essential to mitigate against the damages of cyberattacks. The impact of a cyberattack on one company cannot be contained, because it will inevitably have repercussions across customers, clients and other third parties. Often, detecting the source of a cyberattack demands substantial resources, which makes the damages even more sustained. In addition to reputational damage, the financial costs of cyberattacks are exorbitant.

Cybersecurity and cyber-threat risk mitigation have become significant fields of study and have developed into industries of their own. At UJ, hackathons have been held in order to raise awareness about this, and technology companies, in fact, employ large numbers of professional hackers to test the security of their products or software.

Deploying blockchain technology

Blockchain is a decentralised, encryption technology that secures stored information. A blockchain is essentially a digital ledger or list of transactions that can be shared on a peer-to-peer network. The use of encryption in this ledger, however, prevents users from making alterations unless the majority of the users sign off on them. Every transaction, including all of its information, is stored on many different systems at once across the internet, and this makes it hard for anyone to tamper with, or steal, the data, since they would have to hack all systems in the chain.

To illustrate how this technology works, let us look at the analogy of a number of people in a village collecting their social grants. The first person to collect his R1 000 is Thabo. This is witnessed by his neighbour, Janine, who is also there to collect her grant. In blockchain, the information relating to the transaction of Thabo collecting his R1 000 is called a 'block', and the witness to this, who is also completing a transaction, is called a 'miner'. One transaction verifies the next in blockchain, so if a third person, Thami, also collects his social grant, this transaction will effectively be verified by the other transactions that have been completed before him. The word 'chain' in 'blockchain' denotes all the previous witnesses to each transaction, making it increasingly harder to change the information because to do so you would need to conspire with all the witnesses. However, if we imagine this process unfolding online, we will be closer to understanding blockchain technology. Transactions secured using blockchain are done with the aid of computers, and thus blockchain is a cyber technology.

According to Deloitte, in 2016 alone, over US$1 billion was invested in blockchain by financial services and technology firms globally, and such investments are predicted to increase exponentially over the next five years. According to a 2016 Gartner report, the technology is currently at the peak of a hype cycle. As Ed Powers, Deloitte's US Cyber Risk Services leader, put it, 'While still nascent, there is a promising innovation in blockchain towards helping enterprises tackle immutable Cyber Risk challenges such as digital identities and maintaining data integrity.'[6]

Each block of information in a blockchain is given a digital timestamp, which records when the transaction happened, and what took place. Blockchain was developed in the early 1990s by Stuart Haber and W Scott Stornetta. Initially, the idea was to create a system in which the timestamps on documents could not be tampered with. This was improved on in 1992 when Dave Bayer, Stuart Haber and W Scott Stornetta allowed several document certificates to be collected into one block. Blockchain was then used by Satoshi Nakamoto (a pseudonym) to develop the bitcoin currency. This saw the implementation of timestamp blocks that did not need to be signed by a trusted party.

Currencies, such as the South African rand, are based on the concept of trust. Many years ago, however, the value of all of the money printed in the country was guaranteed by the central bank by means of its gold reserves. Essentially, this means that you could take cash to the central bank and exchange it for its equivalent value in gold. If the government wanted to put more money in circulation, it had to ensure that it held enough gold reserves to guarantee the money. This system was known as the 'Gold Standard'. So, when people exchanged money for goods, they were, in effect, transacting using gold. But when, in 1971, the US ran out of gold to guarantee its printed money, President Richard Nixon stopped linking the US dollar to gold, bringing the use of the Gold Standard to an end. From that point on, people put their faith in the value of a currency simply based on trust in the country that printed it, believing that the money would hold its value and be useful to them, and also trusting that unauthorised people would not be able to counterfeit this money.

Blockchain is essential to cryptocurrencies because the money generated using this technology cannot be easily faked. Cryptocurrencies, such as bitcoin, are so called because the word 'crypto' means that they are created digitally using the principles of encryption. So, Madiba's M-Plan was perhaps a form of crypto-politics – secret support and mobilisation for a political objective. Crypto-politics is such a powerful tool that the Chinese leader Deng Xiaoping gave the Chinese people the following advice on their quest for global power: 'Hide your strength, bide your time.'

Blockchain can be used in many different fields. For example, it can be used in order to secure the medical records of hospital patients. As mentioned in Chapter 3, there is no national database for medical records in South Africa. If I fall sick in Thohoyandou and go to Tshilidzini Hospital, a medical file will be opened for me there, and my medical records will be stored in that file. If I later fall sick in Diepkloof and go to Chris Hani Baragwanath Hospital, a new file – one that does not contain the information stored in my file at Tshilidzini Hospital – will be created. Every time I change hospitals, my new doctor will not have my complete medical records. This, of course, compromises the quality of the medical service I will receive from the doctors.

One of the elements that impede the integration of medical records is the security of information. Blockchain can be used to secure such information so that when the medical records are integrated, a complete, accurate and secure medical history can be generated. Another example of the use of blockchain is in the management of school and university certificates, which can be secured using this technology.

Even though there are many social benefits of using blockchain, the technology also has its downsides. One of these is that it demands significant amounts of energy and involves onerous running costs. This then means that those without financial means can easily be locked out of this technology. Another problem is that blockchain may be vulnerable to coding errors. In our paper, 'Blockchain and Artificial Intelligence', Professor Bo Xing and I indicate that blockchain is susceptible to software slip-ups. We thus propose a framework in which AI can be used to improve the design of blockchain. This can be achieved, for example, by replacing the need for the miners, or witnesses, in the blockchain, to be human beings; instead, we can use AI to perform the verification.

Blockchain is one of the drivers of the 4IR. It is disrupting the world of work. In the field of finance, the auditing process is important in ensuring that investors have access to information about what is really happening in the companies they intend to invest in. However, auditors produce their reports based mainly only on a sample of the activities of the company, which, of course, means that it is always a snapshot and not a complete overview. Deploying blockchain to record transactions will mean that there is always a complete picture of all of the transactions that take place. This is because every transaction in a blockchain is verified by witnesses, or miners, which then removes the need for auditors or accountants.

Another field that will be disrupted by this technology is banking, where blockchain reduces transactional costs, thereby making this industry more efficient. Other areas that will be improved because of blockchain include supply-chain-based sectors, transportation and tax collection. This will greatly benefit, for example, South Africa's tax-collection agency, the South African Revenue Service (SARS).

As a continent, can we implement this technology with the same effectiveness as Madiba's M-Plan? Do we have adequate electricity to support the enormous energy consumption it demands? Are our industries ready to absorb the technology, in particular, and all emerging technologies, in general? Do our institutions of higher education have adequate skills to understand all these emerging technologies? Are universities flexible enough to adapt to the rapid changes happening in the field? The answers to these questions are mixed, but we have no option but to adapt; otherwise, we will be mere spectators and subjects of this revolution.

CHAPTER 11

Social networks

In the last few years, my Twitter, Facebook and Instagram accounts have tailored my newsfeeds broadly in line with my interests and the content I usually engage with. This is based on the accounts I follow, the content I post and the updates that I tend to Like and comment on. For instance, on Facebook, posts from my colleagues about AI usually appear first when I open the app. Some of the recommended content from news sites is about universities, and my Friends suggestions are often people I have come into contact with in recent months. This is a marked change from when Facebook first started. According to market-data collector Statista, there are an estimated 2 billion internet users active on social media and 'these figures are still expected to grow as mobile device usage, and mobile social networks increasingly gain traction'.[1] Facebook remains the most popular social network, with 1.7 billion active users. WhatsApp and Facebook Messenger are in second and third place, with one billion users each.

The genesis of Facebook

In 2003, during his second year at Harvard, Mark Zuckerberg started FaceMash, a website where users could compare the attractiveness of female students. The site presented users with photographs of two female students, placed side by side, for them to decide who was more attractive. These pictures were taken from Harvard's facebooks – a directory of individuals' photographs and names. In its first four hours online, the website attracted 450 visitors and 22 000 photo views. FaceMash was taken down a few days later, but the idea behind it evolved into TheFacebook, a new website and online facebook for Harvard. The intention, at first, was to connect students at the university. Within the first month, half of the undergraduate students had created an account. TheFacebook then expanded to other American universities, and by the time it changed its domain name to Facebook in 2005, it had six million users.

At first, the site was quite rudimentary. Each profile contained a profile picture as well as basic personal information, such as gender, birthdate, an email address and hometown information. There was also a place to put up a status update that would inadvertently start like this, 'Tshilidzi is …' By 2006, a feed was added, which served as a log of one's own activity. You could see what statuses I had put up, if I had changed my profile picture and whether I had written on someone else's Wall. This served as Facebook's model, with a few changes here and there over time. In 2009, the website went mobile and with it came the launch of News Feed, which many at the time scoffed at and compared to Twitter. News Feed provided real-time updates of what people were doing. This was also the moment Facebook began using an algorithm that, at first, offered a sorting order for news, and then advanced to allow you to choose what you wanted to see first. By 2016, the algorithm prioritised content based on how much a user would react to similar content. This development on Facebook substantially changed how brands interact with us. All of a sudden, brands could tailor their content to individuals, as newsfeeds became more personalised. How was this done, and how have social networks changed from a way of keeping in touch to perhaps the best example of leveraging big data? According to a market research report by Credence Research, total global spending on AI in the social media industry is expected to reach around US$33.3 billion by 2027.

The shift towards personalisation

In 2015, Facebook released Facebook Security Check, a feature that allowed users to mark themselves as safe during times of disaster. This brought the world closer together as our reliance on traditional communication to notify family and friends of our safety shifted to social media.

Over the years, Facebook has also refined its ability to personalise content for users. Personalised marketing campaigns are now based on using relatable stories, targeted videos and influential blog posts. It is no wonder then that in the last few years, Facebook has also acquired Instagram and WhatsApp.

But how does Facebook achieve this personalisation? An algorithm ranks posts based on how much it expects users to enjoy them. This is based on ranking signals, or data points from a user's previous interaction with posts. For example, the algorithm will factor in the media preference of the user, such as whether he or she prefers to access video, image or text posts. It will also factor in a user's frequency and level of interaction with different people. If, for example, a user interacts more with a colleague than he or she does with his or her siblings, the colleague's posts will be ranked higher. This also factors in how often users Like a specific person or page's posts. According to Facebook, your News Feed is based on three factors: who you typically interact with, the type of media used in the post, and the post's popularity. Many other social networks similarly tailor content.

Music apps, such as Spotify, Pandora or Apple Music, much like Netflix, are capable of recommending content based on your interests. They monitor the choices you make, insert them into a learning algorithm, and suggest music you are most likely to enjoy. For example, if I regularly listen to Eric Clapton, it may suggest playlists that include Dire Straits.

In 2018, ByteDance, the owner of the wildly popular short-form video app TikTok, surpassed Uber as the world's most valuable technology start-up company with a value of over US$75 billion.[2] This was a remarkable feat when you consider that you would be hard-pressed to find someone who has not heard of the ride-sharing company. TikTok, on the other hand, has a niche market. The interactive app allows you to connect with friends, comment on videos and follow other people on the platform through lip-syncing and dancing videos. It has been a catapult for influencers – an influencer is a person who establishes credibility in a specific area, such as travel, fashion or fitness, and who garners large audiences – who create specialised content through the app for brands. In Australia, Aboriginal teenagers have also, for instance, used the app to create awareness around racial disparities.

What is remarkable is that TikTok does not yet allow you to monetise your content on the app as you would on YouTube or Instagram, but it has provided a platform for creators to secure sponsorships and brand deals for individual posts. ByteDance's flagship product is Chinese app Jinri Toutiao – which means 'Today's Headlines' – a popular news-aggregation service that uses AI to track reader habits and push them stories from various sources, hooking millions of readers by personalising the selection of articles.[3]

Social media platforms, such as Instagram or Snapchat, also use AI. According to a survey conducted by investment bank Piper Jaffray, 90% of teens report using Instagram at least once a month, 70% say that they prefer brands to contact them about new products through Instagram, with Snapchat following as the preferred method for brand engagement at about 50%. Snapchat is the favoured social media channel, with 41% of those surveyed naming it as their favourite, compared to 35% for Instagram. Just 6% of teens named Facebook as a favourite.[4]

Here, AI and augmented reality (AR) features have become commonplace. For example, on Instagram, some of the most-used filters include those that allow users to superimpose Louis Vuitton signs on their faces, to digitally etch freckles onto their cheeks, to enlarge their lips, and to have pink dollar signs floating around. While this may seem like another Instagram gimmick, something that draws more users to the app, it is, in fact, premised on 4IR technology. Interestingly, these filters are created by designers and other third parties as opposed to Instagram itself. In May 2019, Instagram launched a beta program that allowed people to create custom face filters and spread them to their followers on Instagram Stories through Facebook's Spark AR Studio.

How does this work? First, it is key to understand AR, which is a technology that superimposes a computer-generated image on a user's view of the world around them or on a picture, producing a composite. In this way, the user's perspective of real-life objects is enhanced, for example by supplementary computer-generated information about the real world or, in the case of Instagram AR filters, images and dynamic content, such as floating dollar signs superimposed on selfie images.

Another example of this, on Instagram, is the new AR shopping feature, which allows consumers to try on products digitally before buying them. For example, you could try on the latest shade of Mac lipstick on your phone, to see how you would look. Of course, this is all in a virtual space. Some home-decoration websites also have apps that can be used to visualise possible permutations.

In recent years, there has been a rise of social media influencers. Instagram has used AI to introduce virtual influencers. Miquela is an AI influencer with 2.3 million followers. Just like any other influencer, her posts are perfectly planned, and she has a themed feed, has sponsored content, and gives her followers useful advice and brand recommendations.[5] But she does not actually exist – rather, she is run with AI technology. This has not stopped her career from taking off. In 2020, she worked with Prada by posting 3D-generated GIFs of herself at the Milan Fashion Show wearing the spring/summer 2018 fashion collection.[6] On Prada's Instagram account, she gave their followers a mini-tour of the space, just as any influencer would for a particular brand.[7]

But Miquela is not an outlier – there are many more like her. Luxury fashion house Balmain recently announced a Balmain Army made up entirely of computer-generated imagery (CGI) models. There's even a dedicated modelling agency, The Diigitals, for digital models.

AI is also being used to help limit negative interactions online. Instagram has started using machine-learning algorithms to detect and remove mean comments and posts in a bid to curb cyberbullying. User complaints about posts that may amount to bullying or malicious behaviour are now also managed by AI, which queues options and files complaints by order of intensity or severity.

How are social networks transforming different sectors and impacting us?

The world of academia and research has embraced social networking as an invaluable form of promoting research, disseminating knowledge and collaborating. Online tools can be used to connect and collaborate with other scientists. Currently, for academics, Research Gate and academia.edu are by far the most utilised social networking platforms for scientists. While some of these social networking sites for academics are commercialised, others provide invaluable academic depth and connectivity. It is now accepted practice for academics to use social media to increase research impact. Platforms such as ORCID or Mendeley, along with more well-known platforms, such as YouTube or Twitter, allow more extensive dissemination of research, creating audiences beyond conventional readers. For the higher education and research communities, this is a significant breakthrough because critical research was often relegated to a narrow, parochial audience in the past.

At UJ, debates hosted on Cloud-based platforms have brought together experts and ordinary folk to talk about critical issues that impact on society. Generally, academia has always been accused of operating in closed circles and having ivory-tower mentalities, but the arrival of the worldwide web and social media platforms have significantly widened the audience for experts who wish to disseminate their theories and research.

What opportunities are there for academia to research the impact of social networking on individuals? An interesting area for research is the impact of social media on the way people project themselves. Often, there is a dissonance between the projected self and the real self. The notion has received attention, especially in the case of high-profile personalities posting happy pictures or status updates, only to take their own lives shortly thereafter. Another potential area of research is to explore the relationship between people having large social networks and their psychological well-being, with many reporting feeling lonely in spite of their online popularity. In this respect, there are opportunities for researching the adverse effects of social media, including the fostering of feelings of isolation and alienation among users.

Social networks have also completely transformed the media industry. Traditional newsprint has had to take a backseat because social media platforms more readily satisfy an immediacy that is craved by audiences. This has seen newspapers reducing actual print runs, while accelerating the posting of live news on multiple platforms. Using paywalls as a monetising tool, newspapers are all-but digital, although many media houses have come to understand that using paywalls potentially restricts their reach. With the fast-paced nature of news cycles, articles behind a paywall can become rapidly redundant, thus reducing the impact of a particular story. Furthermore, media houses have had to deal with the problem of people copying and sharing their content online. These problems have forced media houses to reassess their traditional revenue models.

Social networking has also impacted the way that customers pose queries or post complaints about companies. This communication is now channelled through various web and social media platforms. For example, it is much easier to submit a query to a bank by filling out a form on its website or chatting to an agent online than it is to have to physically go into a bank.

The 4IR makes it almost mandatory to have a robust online presence, and a lack of engagement on social media could result in severe customer aversion. Often, when they have experienced poor service, customers take to Facebook or Twitter to blame, shame and elicit a resolution to a problem from a company. The time that it takes for a company to respond could be detrimental to their business – especially if it is perceived to be too slow, for instance.

Today, companies rely heavily on social media presence to enhance or create brands. They also use social media as a primary or secondary source of data for business analytics in order to inform their business decisions and strategies.

Companies can also use social media to promote sales and specials. Online store Superbalist gives users discount codes on its Instagram account, while stores such as Woolworths and YDE, for instance, may inform users about seasonal sales via Facebook or Twitter.

Musicians and movie stars use social media to promote forthcoming attractions in a previously inconceivable way. A few years ago, *The Hunger Games* trilogy was promoted via a dedicated, official Instagram account. By sharing behind-the-scenes shots, graphics from the movie, interviews and materials from other campaigns involving its stars, the account garnered 1.4 million followers on the platform.

Social media has also fundamentally changed the way people express their opinions on topics and rally around causes. In the past, in order to make their opinions known publicly, people had to write letters to newspapers or magazines. Now, social media allows people the freedom to express these views without being filtered, irreverent as they may be, and to rant publicly.

Social media thus has the power to ignite the public and unify them in a movement. People used this power of social media, for example, in the Arab Spring, gaining traction for and organising the movement through Facebook. In 2007, the Kenyan website Ushahidi was used to track post-election violence in that country and, closer to home, in South Africa, the #FeesMustFall movement gained considerable traction on Twitter and Facebook.

The 2019 WEF in Africa, for instance, coincided with a tragic incident of gender-based violence that had South Africa in an uproar. Mass protests and calls for action from leaders followed the rape and murder of Uyinene Mrwetyana, a University of Cape Town student, and the murder of a young boxing champion, Leighandre 'Baby Lee' Jegels. The jury is out as to whether the official response from the president was genuine or merely an obligatory one necessitated by the outrage expressed on multiple social media platforms.

In 2012, there was a social media frenzy around Ugandan militia leader and war criminal, Joseph Kony. The Kony 2012 video, which was released with the aim of making the atrocities known globally in order to have Kony arrested, has over 100 million views and almost 1.5 million Likes on YouTube. In 2012, when it was first released, it was YouTube's most-liked video, and it is said to be the first-ever to reach one million Likes. While there was no tangible outcome and Kony has become somewhat of a meme, the incident showed how a movement can quickly become viral – and how quickly people lose interest.

Social media is fundamentally changing the way we interact with each other and with industry. A tweet can be sent out in an instant to millions of users. New product releases have more range and scope than ever before, and brands can engage with their consumers in more tangible ways. Not only is there a more substantial following of and interest in brands, but there is also a greater immediacy to responses. Brands now have to be online all the time in order to keep abreast of what is going on with their consumers. Some brands use real-life events that are in the public interest to amplify their reach. Nando's, for example, often uses events such as SONA or the annual budget speech made in parliament as a basis for adverts characterised by the use of witty puns.

Another downfall of the proliferation of social media is that these platforms enable people to take liberties online that they would not in person, fostering keyboard warriors who go on the attack, pushing their own agenda. While in some instances, this has had positive spin-offs and brought about much-needed changes in society, in other cases, it has been akin to mob justice. Likes, Comments and Shares can often create a frenzy characteristic of a mob, but, on a more positive note, the injustices of society can be surfaced far more readily using the same social media platforms to sway public opinion. In fact, the freedom that comes with social media platforms enables those who would not generally voice their opinion to add to the momentum of a debate.

What is clear is that our interactions are becoming increasingly simple – a matter of typing 280 characters or less and sending it out to followers on multiple platforms. This is particularly relevant to Africa, which boasts the fastest growth rates in internet penetration – jumping by more than 20% in 2018 compared to 2017. This has largely been driven by greater access to mobile devices and data plans.

CHAPTER 12

5G technology

It was in July 2019 that the Minister of Communications and Digital Technologies in South Africa, Stella Ndabeni-Abrahams, issued policy direction on the allocation of the frequency spectrum. The frequency spectrum is what telecommunications companies need to make it possible for people to communicate via a cellular phone. For example, a telecommunications company requires spectrum for its customers to communicate and download content from the internet. Frequency spectrum, like any other commodity, is a scarce resource. This was the first major allocation of the spectrum since the 2.1-gigahertz (GHz) band that helped operators in their 3G network deployment in the mid-2000s. Vodacom and MTN, Africa's two biggest telecommunications companies, were allocated this spectrum by the South African government in 2004 and 2005, respectively, while Cell C received this spectrum in 2011.[1] As Ndabeni-Abrahams explained at a ministerial briefing in 2019, 'We delayed so much with 4G. I'm committing to one thing: we are not going to delay with 5G. But we are not going to allow what happened in the past to happen with 5G, as much as we're not going to delay. We've got to make sure we use that opportunity to transform South Africa and the industry and give people an opportunity to participate.'[2]

Simply put, fifth-generation (5G) is a telecommunications technology that is sufficiently localised that it can transmit information very fast. Instead of only having some telecommunications towers located in areas in order to serve many people, as is the case with older generations of technology, such as 4G, in the delivery of 5G, hubs are located near users, such as on street poles, to serve fewer users and thus facilitate the fast transmission of data. The availability of the frequency spectrum will facilitate the introduction of 5G and associated technologies. The frequency spectrum is, in essence, the new gold standard of telecommunications companies. 5G communication – which combines ubiquity, reliability, scalability and cost-efficiency – is a crucial driver for the introduction of the IoT. Currently, the US and China are battling it out to see who will dominate 5G technology. How this war will end is a subject for debate, and different investors are putting their money on whom they think will win.

Frequency spectrum

So, what is the frequency spectrum? To understand the frequency spectrum, we need to understand frequency. Mobile or cellular communication is achieved using electromagnetic waves. Electromagnetic waves emanate from the same branch of physics that unites electrical and magnetic forces, and which gave us electricity and the electric motor. The electric motor is used to power electric cars, fridges and assembly lines. Assembly lines gave us mass production of goods and services, while electromagnetism also gave us Einstein's theory of relativity as well as the means of communicating via cellular phones.

Frequency is the number of waves transmitted in one second, and it is measured in hertz (Hz), named for Heinrich Hertz, the German scientist who demonstrated the existence of electromagnetic waves. Radio broadcasting uses one frequency to broadcast to many people. For instance, Power FM broadcasts at a frequency of 98.7 MHz, so the transmission equipment that Power FM uses communicates 98.7 million electromagnetic waves every second. The receiver in our radios will capture this signal by tuning at 98.7 MHz.

Telecommunications companies allocate a dedicated channel of communication to users. If Thendo calls Denga, the two need a dedicated channel to communicate so that no third party can hear their communication. So, because telecommunications companies have multiple users, they need multiple frequencies (or a frequency band or spectrum) – rather than one frequency, as is the case in a radio broadcast. In this regard, telecommunications companies require frequency bands, such as those within which the Global System for Mobile Communications (GSM) standard operates (for example, the 900 MHz band).

Within the constraints of scarcity, the frequency spectrum needs to be allocated in such a way that its usage benefits South African society. But what are the essential considerations required for the allocation of the spectrum? Firstly, the allocation of spectrum should facilitate investment. There is also no point in allocating spectrum to a company that has no expertise in the deployment of telecommunications infrastructure. If we allocate spectrum to such entities, we will end up with a rental market in which entities acquire spectrum licences simply to sell it on, thus harming consumers because operators inevitably pass those costs on to consumers.

The second consideration is that the allocation of the spectrum must facilitate competition, and in order to ensure that, we need more players in the market. In South Africa, some 400 telecommunications players who require spectrum cannot access it. More competition leads to lower costs of data, so the mantra #DataCostsMustFall cannot be achieved outside the promotion of competition. South Africa can only reach the 4IR if the 5G platform is enabled and in a cost-effective manner.

The third consideration is that spectrum allocation should also facilitate the entry of new players, especially small- and medium-sized enterprises. In his conception of the principles of evolution, the revered scientist Charles Darwin observed that systems that do not allow for new players could not survive. In this regard, if the South African telecommunications market does not allow for new players, it too shall die.

In South Africa, to ensure that the new policy direction meets the attributes for an efficient telecommunications industry, the Department of Communications and Digital Technologies contracted the CSIR to develop a framework for the allocation of spectrum. This framework uses the 80/20 principle, in which 80% is allocated towards functionality and 20% is for industry transformation considerations. We can have a competitive and transformed telecommunications industry if the public–private skills balance is adequate. The private-sector exploitation of the public sector, a situation dominant in many African countries, can be eliminated if we follow the guidelines and educate our public-sector officials.

The race to 5G

In recent times, the race to lead the 4IR has begun to play itself in trade tensions between the world's two largest economies. Not only has the US put some Chinese AI companies, such as Huawei, on sanctions lists, but the export of American AI software to China has also been restricted. Over two years, this has impacted markets and stunted global growth. In an article in early 2020, CNBC tracked back to the heart of the matter, indicating that the

Trump administration has long maintained that Huawei is a national security threat and has tried to convince other nations not to use its equipment for 5G. Yet, the US does not have any policy around 5G or a viable alternative. As CNBC journalist Arjun Kharpal asserts, the crux of the matter is that '[any] attempt to try to set up a Huawei rival in 5G is just too late'.[3]

South Africa is, however, trailing significantly behind. In March 2020, following three months of public submissions, the Independent Communications Authority of South Africa (ICASA) announced that it would begin its spectrum-licensing process in April. At the time, ICASA announced an emergency release of broadband spectrum to meet a spike in internet demand during the national lockdown following the outbreak of the COVID-19 pandemic. This did not, however, interrupt the licensing process, which is expected to be concluded by the end of 2020. ICASA is anticipated to procure the services of an expert spectrum auctioneer to manage the licensing process that will allocate spectrum to commercial operators. This is an exciting process. In the past, South Africa did not formally license spectrum suitable for the rollout of 4G networks, and this forced mobile operators to reallocate existing spectrum assets used to build their 2G and 3G networks.[4] ICASA's spectrum auction is expected to help operators build better coverage and expand the reach of their networks.[5]

President Cyril Ramaphosa also vowed in his SONA earlier in February 2020 that the licensing of high-demand spectrum will be concluded by the end of the year. As President Ramaphosa put it, 'An important condition for the success of our digital economy is the availability of high-demand spectrum to expand broadband access and reliability.' There is an urgency because, not only does South Africa need to see frequency spectrum rolled out, it also has to consider its data costs. The world is entering an era in which data is becoming as essential to us as many of our other basic needs. In March 2020, MTN and Vodacom, which between them control about 70% of South Africa's mobile industry, were given an additional month to establish a deal with the South African Competition Commission that would result in the reduction of data costs.

In December 2019, the Competition Commission released the final report of its *Data Services Market Inquiry*, which recommended that 'the two mobile network operators reach an agreement to reduce data prices, particularly for monthly bundles, and to address the structure of data pricing, reducing the cost per MB for smaller sub-1GB bundles relative to the 1GB price'.[6] As Trade and Industry Minister Ebrahim Patel put it, 'The prices are higher than they should be and higher than any other markets elsewhere in the world. Data is reshaping the 21st century. If we want to grow the economy, we need to have the lowest possible data prices.'[7]

The decision has already had an impact. In March 2020, Vodacom announced that it would cut data costs by 30%. MTN followed suit a week later, announcing that it would reduce the price of its monthly bundles of 1GB and below by between 25% and 50%.[8]

But 5G is not merely a marketing tool for telecommunications companies to stay competitive. The world is increasingly going mobile and consuming more data. We can do our banking, reply to emails and google the latest on 5G in a matter of seconds. Yet, the current frequency spectrum is also becoming increasingly congested and simply cannot keep up with the demand. This is an echo of what we have experienced during load-shedding, for example. As more people in the same area try to access online mobile services at the same time, there will be gaps in connectivity. At concerts, you may struggle to upload a video to Facebook or Instagram or even send out a simple WhatsApp because everyone at the stadium is trying to do the same. But 5G is much more effective in handling thousands of devices simultaneously. This does not merely encompass mobile devices either but includes video cameras and smart street lights, for instance.

The history of 5G

What is 5G, and how have we reached this point? Although South Korea was the first to offer 5G, the technology is expected to be adopted by most countries and will drive the IoT and big data. As vice-president of Ericsson for the Middle East and Africa, Nicolas Blixell, stated at AfricaCom – the largest gathering of the tech community on the continent – in Cape Town in 2019, 'If you don't have 5G, you don't have the fourth industrial revolution.'[9] Each generation of wireless standard, from 1G to 5G, has had an increase in data-carrying capacity, and US mobile network operator Verizon estimates that 5G will be between 30 to 50 times faster than 4G.

But understanding the history of 5G is essential. Much like the 4IR, we have to appreciate what has come before for us to know where we are heading.

The first generation of mobile networks – or 1G – was launched by Nippon Telegraph and Telephone (NTT) in Tokyo in 1979. By 1984, it was rolled out to the whole of Japan.[10] This, of course, was almost 30 years before the use of data as we understand it now. That first generation was an analogue telecommunications standard, but coverage and sound quality were low. There was no roaming support between various operators and, as different systems operated on different frequency ranges, there was no compatibility between them.[11] This, of course, was a world away from watching cat videos on YouTube or streaming live videos via Facebook.

The second generation (2G) was a slight upgrade on 1G. Here, calls could be encrypted, and digital voice calls were significantly clearer, with less static and background crackling.[12] This saw the advent of the short-messaging service (SMS) and the multimedia-messaging service (MMS). While these innovations may seem rudimentary now, they nevertheless paved the way for the apps we use today. Suddenly, instead of calling, I could simply text a colleague that I would be running late; I could text someone to explain that I was in a meeting and so could not answer their call. Companies would often include slideshows in MMSs as a form of advertising. This also saw the launch of online instant-messaging services, such as Mxit and mig33.

In 2001, 3G was launched. For the first time, users could access data from any location in the world. There was an increase in data-transfer capabilities, with 3G being four times faster than 2G, and this gave rise to video conferencing and streaming, for example.[13] Suddenly, you could use Skype on your phone. This development coincided with the phenomenal growth of BlackBerry and the rise in popularity of its messaging service, BBM. In 2007, the first iPhone was released.

Although we are on the cusp of 5G, we still mostly use 4G, which has become synonymous with streaming. It is now easier than ever to pull up a YouTube video on your phone or to watch stories on Instagram. Of course, this has called for more sophisticated technology. We could hardly imagine 4G capabilities on the old, 3G version of the Nokia 3310. While the shift from 2G to 3G called for a switch in SIM cards, devices had to be specifically designed to support 4G. To consider what a jump this has been, try to connect to the internet when your phone says '3G' next to the network bar. You may find that connectivity is far more sluggish than you are used to.

Understanding 5G capabilities

Interestingly, we are now making the leap towards 5G even though 4G coverage remains low in some areas. Of course, 5G smartphones give you the capability to download a movie in less than a minute, stream videos much like you stream audio, and complete faster web searches, but it is much more than that; 5G can, for instance, support self-driving cars. These are vehicles that are not driven by a human driver, but instead are autonomous, using sensors to perceive their surroundings and a control system (which we looked at in Chapter 5) to interpret this data and identify routes. So, while Uber has been making huge investments in developing self-driving car technology, it is 5G that can make this a reality.

Another benefit of 5G is that it can also transform health care, particularly in remote areas. Doctors could use robots to operate on a patient from a thousand kilometres away, and people could have access to specialists, even those from a different country. Gamers could also have more immersive video-game experiences through AR or virtual reality (VR). While there are such capabilities now, there would be no lag, and the experience would be more realistic. Concert-goers could go backstage to meet their favourite artists through AR or VR. Video conferencing could become more reliable, with a better and faster connection, one that is not intermittent and does not have lag. These are just some instances of the capabilities of 5G.

There are, of course, economic considerations as well. In 2019, IHS Markit forecast that 5G could generate US$13.2 trillion in sales enablement by 2035. McKinsey Global Institute estimates that by the end of this decade, 'global GDP could increase by $1.2 trillion to $2 trillion by implementing new use cases across four domains alone: mobility, healthcare, manufacturing and retail'.[14] By 2030, the research firm forecasts that '5G deployments using millimetre wave spectrum will cover just 25% of the global population, at an estimated cost of $700 billion to $900 billion'.[15]

Yet, all of this will, of course, remain in the hands of countries that are already at the forefront of 5G and will not likely be as effective in rural areas. Around 80% of those in urban areas will be covered with high-band 5G (the fastest form of 5G) in 2030, compared to 'almost none' in rural areas. Also, coverage will be higher in countries that are leaders of 5G technology than in countries that are not. As McKinsey put it, 'Due to high investment costs, high-band 5G networks are likely to be limited to select geographies for the foreseeable future.' Low- and mid-band 5G, which will perform similarly to 4G, is expected to reach 80% of the global population at the same time, costing between US$400 billion and US$500 billion.[16]

South Africa's 5G landscape

So, where does South Africa stand? While we wait in anticipation for ICASA to get the ball rolling, there have been some moves towards implementing 5G capabilities. In 2019, data-only mobile operator Rain launched the first commercial, fixed-wireless 5G network in parts of Johannesburg and Tshwane. However, while this has delivered faster internet, it has been coupled with connectivity issues and outages. In early 2020, Vodacom announced that the telecommunications company expects to offer 5G mobile services in South Africa before the end of 2020, thanks to a recent roaming agreement with local data provider Liquid Telecom.

This came just after Liquid Telecom announced that it would launch the first '5G wholesale roaming service in South Africa in early 2020', using its share of the 3.5 GHz (one gigahertz is equal to 1 000 million hertz) spectrum and 'allowing mobile network operators to have open access to the new network.'[17] This could be a reality sooner than we anticipate as Vodacom has already launched 5G in Lesotho. Yet, this rollout will depend on how fast ICASA can complete its task. Many of South Africa's ambitious 5G plans rely on spectrum allocation.

As Chuck Robbins, CEO of Cisco Systems, put it, 'The opportunity we have is to build a secure, intelligent platform that solves some of the world's greatest problems at scale. That is what is possible with hundreds of billions of connections and the capabilities that we can deliver together.' During his SONA in 2020, President Ramaphosa outlined plans for a new smart city in Lanseria, which will be home to up to 500 000 people over the next decade. This is of significance when you consider that South Africans migrate to urban areas for job opportunities and that it has been estimated that by 2030 about 70% of South Africans will live in urban areas. As President Ramaphosa explained, the new city will be 5G-ready and aims to be an international exemplar for green infrastructure. This, however, will remain little more than a promise if we do not see the swift rollout of 5G.

PART 3

Business

Perhaps the most significant overhaul we have seen in the 4IR is of the financial sector. Where automation once seemed to impact only blue-collar jobs, there is already evidence that white-collar jobs are not only impacted, but also at risk. As these changes begin to unfold in banking and taxation, we must embrace technology while also preparing our workforce for it. This section looks at the impact of the 4IR on the financial services sector and delves into the benefit of creating greater economic growth, platforms for trade and more efficient markets.

CHAPTER 13

Economics

At the start of 2020, Stats SA released its GDP figures for 2019. It turned out that South Africa had not only fallen into a recession – defined as two consecutive quarters of negative growth – in the second half of the year, but also that growth for the year was at only 0.2%, a downturn that had been speculated about for quite a while.

This figure was even lower than the South African Reserve Bank's growth estimate of 0.4%, which it had revised down, and only slightly above the estimate of 0.3% in Finance Minister Tito Mboweni's budget policy statement delivered just a week before this grim prognosis. It's a particularly frightening number when you consider that unemployment at the time was inching towards the 30% mark. This did not even encompass South Africa's discouraged workers, who had given up on finding work, which, if included, would put the estimate closer to the 40% mark. This, of course, was partly based on long-term weakness in most sectors, a decline in public-sector investment, low export growth and extremely modest growth in government and household consumption. Yet, there was another blow – this time delivered by Eskom. The return of widespread load-shedding in 2019 resulted in seven out of ten industries contracting in the fourth quarter.

There has been a chorus of voices putting pressure on the government to turn the economy around. With low growth already forecast based on slow economic activity and the return of load-shedding in 2020, the economy was dealt a further blow by the coronavirus, which necessitated a national lockdown. The Reserve Bank forecast in April that the economy could contract by as much as 6.1% and anticipated that the country's budget deficit could exceed 10% – a far cry from the 0.9% that National Treasury had anticipated in the February budget.

In a statement, following a visit to the country in 2019, the International Monetary Fund (IMF) said, 'South Africa faces a prolonged period of weak economic growth marked by rising unemployment, inequality and greater credit-rating risk if the government does not act fast to implement reforms.'

This followed a series of warnings from rating agencies Moody's Investors Service and S&P Global Ratings, which had both changed their outlook on South Africa's debt to negative, sounding another warning that it was time to act. As the IMF put it, 'South Africa has to create an environment conducive for private sector investment and take a decisive approach to implement structural reforms to boost economic growth.' Since then, South Africa's credit rating has been downgraded to junk status by all three major credit-rating agencies.

This all but echoed the clarion call I made to the South African government at the end of 2019 to act then or to risk being left behind, which would place the economy in an even more precarious situation. But why is this important? It is vital because an economically, intellectually and technologically developing country cannot be free. As former UN Secretary-General Ban Ki-moon put it, 'Saving our planet, lifting people out of poverty, advancing economic growth ... these are the same fight. We must connect the dots between climate change, water scarcity, energy shortages, global health, food security and women's empowerment. Solutions to one problem must be solutions for all.'

What figure should we be looking at for growth?

According to the National Development Plan (NDP) – which is, in essence, the country's economic blueprint – growth would need to be about 5% to make so much as a dent in the rising unemployment rate. Yet, the economy has not grown by more than 2% annually since 2013 and is still struggling to increase momentum, notwithstanding political changes and efforts to lure investment into the country. This also came as South Africa struggled to hold on to its final investment-grade rating, from Moody's Investors Service – and, of course, these fears were realised on the day the national lockdown began, as Moody's dealt this final blow, citing the deterioration in South Africa's fiscal strength and structurally low growth as reasons for the downgrade. Following the Reserve Bank's interest-rate announcement for 2019, the credit rating agency stated that the country's high unemployment, income inequality, and social and political challenges had proved to be substantial obstacles to the government's plans to increase growth and contain fiscal deficits.

As the Reserve Bank governor, Lesetja Kganyago, put it, there is an urgent need to implement 'prudent macroeconomic policies and structural reforms that lower costs and increase investment, potential growth and job creation'.[1] Part of that, I would argue, is for South Africa to stay ahead of the curve with the 4IR.

The 4IR and emerging economies

The road to be taken in this context is one that affords emerging economies the prospect of leapfrogging in order to become transformational. One argument suggests that development has to be linear and sequential. However, leapfrogging is what emerging economies do best. This entails a quick jump in economic development through harnessing technology. This strategy requires consensus among government, the private sector and citizens, and it can enable development similar to the boom in the Southeast Asian countries that opted to tap into manufacturing. Fears of job losses resulting from automation already pervade different sectors of society, the economy and the political sphere. Discovering the optimal solution to potential job displacements requires a reorientation of our approach to education, science and innovation. Inequities and inequality in communities can be overcome by levelling the playing field to ensure that technological inequality does not become a new exclusionary barrier.

While some would have us believe that the 4IR spells doom, it could be the key to finding solutions to some of our most deep-seated problems, even as we acknowledge its potential to exacerbate poverty and inequalities. However, if we are to heed its call, we may be able to subvert the negatives that appear to come with the territory. Let us take the example of Rwanda, where the government collaborated with US start-up Zipline to deliver blood supplies by drone to remote areas, thus serving as a model to address the problem of delivering emergency medical supplies to remote locations. Whereas such a delivery would previously take three hours to arrive by vehicle, it only takes six minutes to arrive by drone.

Research can thus be redirected to exploring meaningful and innovative solutions to real-world problems.

The impact of the 4IR on South Africa's economy

Amid this extremely turbulent economic background, there has been much fear around the 4IR. The worry is that the massive job losses that will stem from another industrial revolution will be painful to curb – this is of particular concern in an economy that is not growing or creating any scope for employment, and it is further exacerbated by the social and economic impact of the potentially crippling effects of the national lockdown.

As we saw in Chapter 1, a 2018 Accenture study suggested that up to six million jobs, including both blue- and white-collar jobs, will be lost to the 4IR by 2025. Interestingly, a study published by the Brookings Institution, a US research group, found that an upsurge of AI poses a risk to college-educated labour as technology becomes more and more advanced and extensively implemented by a range of industries – in fact, the researchers say, AI is five times as likely to displace college graduates than those without a degree.

It is clear then that the 4IR has the potential to grow our economy exponentially, but in order for us to reap the benefits, we must understand what it is. A firm commitment to the 4IR is not merely hype – as a lot of current rhetoric would have us believe – but indeed necessary for us to realise our economic fortunes. Unlike in the cases of the previous industrial revolutions, we cannot afford to be left behind. In September 2019, it emerged from the WEF on Africa held in Cape Town that the 4IR may actually accelerate socio-economic development across Africa.

According to a 2018 report by McKinsey Global Institute on the impact of AI on the world economy, 'there is the potential to incrementally add 16% or around $13 trillion by 2030 to current global economic output – an annual average contribution to productivity growth of about 1.2% over the next decade'.[2]

When we disaggregate this, by 2030 the automation of labour could contribute up to 11%, or around US$9 trillion, to global GDP, while innovations in products and services could grow GDP by about 7% or approximately US$6 trillion.[3]

According to the report, 'the introduction of steam engines during the 1800s boosted labour productivity by an estimated 0.3% a year, the impact from robots during the 1990s around 0.4%, and the spread of IT during the 2000s 0.6%'.[4] So, the 4IR, therefore, has the potential to counter slow economic growth and lagging productivity. Importantly, South Africa now has to position itself in such a way that we are ready for this transformation. We risk falling further behind if we do not.

According to researcher Irving Wladawsky-Berger, 'The McKinsey report is based on simulation models of the impact of AI at the country, sector, company and worker levels. It looked at the adoption of five broad categories of AI technologies: computer vision, natural language, virtual assistants, robotic process automation, and advanced machine learning. Data sources included survey data from approximately 3 000 firms in fourteen different sectors and economic data from several organisations, including the United Nations, the World Bank and the World Economic Forum.'[5]

According to Accenture, their 'research on the impact of AI in twelve developed economies reveals that AI could double annual economic growth rates in 2035 and could increase labour productivity by up to 40%, enabling people to make more efficient use of their time.'[6] Similarly, a report from PwC estimates that AI advances will increase global GDP by up to 14% by 2030, the equivalent of an additional US$15.7-trillion contribution to the world's economy. Around US$6.6 trillion of this will come from productivity gains, especially in the near term. These include the continued automation of routine tasks and the development of increasingly sophisticated tools to augment human capabilities. According to the report, increased consumer demand for AI-enhanced offerings will overtake productivity gains and result in an additional US$9.1 trillion of GDP growth by 2030.

All of this, however, is expected to have a lesser impact on Africa. According to PwC, Africa, Oceania and some less-developed Asian markets will see US$1.2 trillion or 5.6% GDP growth. This is significantly less when compared to China, which is expected to see the most significant economic gains from AI, with a US$7-trillion or 26% boost to GDP growth, or North America, which is expected to see financial benefits of US$3.7 trillion or 14.5% of GDP growth by 2030.

The future of economics

It is not just the economic potential of the 4IR that we should be focusing on – developments in the 4IR can also transform the study of economics and allow us to improve economic forecasting. Economists have long been criticised for getting their forecasts wrong, with economic models picked apart for their inaccuracies. According to Prakash Loungani of the IMF, economists have failed to predict 148 of the past 150 recessions. But this is not just a global anomaly. Many economists did not see the interest rate cut at the end of 2019 coming. Headlines are often littered with news of surprise contractions or grimmer-than-expected unemployment figures. And it is not just the predictions that are off. In fact, actual data are often inaccurate as well. As Sunita Menon observed, 'On 6 June 2017, Stats SA's quarterly GDP update made for grim news. Two consecutive quarters of negative growth had pushed the economy into a technical recession. Nine months later, however, the national statistics agency revised its numbers and – just like that – the recession had never happened.' This has become somewhat of a pattern. As more information becomes available, Stats SA has steadily gone back and revised many of its figures. GDP, the calculated value of all the goods and services produced in a country over a specified period, usually a year, is a vital economic metric used by policymakers, economists and investors to estimate the health of the economy. The figure is used to determine how an economy has grown or contracted over time – and thus informs whether the policy should boost or rein in spending, for example. It also allows for a comparison between economies.

There are many obstructions to efficient economic forecasting. As Adam Shaw once wrote for *The Guardian*, 'Increased complexity is not the only problem – forecasts are also made less trustworthy because of a feedback loop. So, if a meteorologist says it will rain, the fact that you take an umbrella out with you does not affect the weather. But if an economist forecasts that inflation will rise by 3% and we react by asking for at least a 3% rise in wages, we have changed the basis on which the forecast was made. Inflation is now likely to rise by more than 3%. The fact that the forecast exists changes the reality it is trying to predict.'[7] Herein lies the potential of AI when it comes to economics. Economic models based on AI will eliminate some of these challenges. By analysing patterns of behavioural economics, AI will help economists make more accurate forecasts. As Walter O'Brien, CEO of Scorpion Computer Services, said, 'Humans have 3% human error, and a lot of companies can't afford to be wrong 3% of the time anymore, so we close that 3% gap with some of the technologies. The AI we've developed doesn't make mistakes.'

In South Africa, historically low levels of business and consumer confidence have kept the economy in the doldrums. In the US, investment bank JPMorgan already makes use of an algorithm that tracks the impact of President Donald Trump's tweets on financial markets. With more precise mechanisms in place to make projections, fiscal and monetary policy will be sounder. For example, if central banks know that a recession may hit, they will be more effective, and their response will be more rapid in enacting monetary and fiscal tools, thereby mitigating the effects of business cycles. In a 2017 paper titled 'Revisions to South Africa's Gross Domestic Product', Asanda Fotoyi, an economist with the non-profit research organisation, Trade & Industrial Policy Strategies, stated, 'GDP estimates … are needed for the monetary policy committee to set interest rates and the National Treasury to set budget limits, but because of the need for timely data, policy decisions are often based on preliminary estimates, which are later revised as more comprehensive data becomes available.'[8] With the use of AI, these predictions and estimations become more accurate, and there is less room for error when it comes to macroeconomic policies. As many proponents of AI will tell you, these kinds of changes will bring economics closer to a science.

CHAPTER 14

Banking

I did not visit a bank branch once in 2019, simply because I can now conduct all my transactions using the banking app on my phone. Banking, something that was once done exclusively at branches, is becoming increasingly digital. These changes, however, require a new breed of tech-savvy leaders who understand technology. Should these new bankers be accountants or engineers or both? When I was completing my doctorate in AI at the University of Cambridge twenty years ago, many of my Engineering classmates became bankers, and today some are even running large Wall Street financial institutions. Modern bankers need to understand both finance and technology, and therefore must be both engineer and economist. The advances in AI are revolutionising banking and deepening digitisation. It is making digital banking intelligent and more efficient for banks in terms of staff count. This digitisation of the banking industry is driven by financial technology (fintech).

The fintech landscape

In 2019, the Reserve Bank launched a new fintech programme to assess the appropriateness of policies and regulatory regimes in light of fintech innovation. This was done in collaboration with the Intergovernmental Fintech Working Group as well as with other policymakers and regulators. In addressing this effort, Reserve Bank governor Lesetja Kganyago said, 'In an increasingly complex domain, collaboration and joint sense-making will remain important.'[1] This, of course, exists within a fraught context of income inequality, unemployment and skills shortages. The income inequality of South Africa's population is high, and many people earn income from the informal and primarily cash economy. As the Centre of Excellence in Financial Services stated in a report, 'The country's significant potential for digital innovation must be considered alongside concerns of whether this will be exclusionary, and whether the transformation will enhance or diminish domestic value creation.'[2] The report suggests that innovation in the space caters to a 'niche, relatively affluent and financially savvy consumer market'.[3] Despite the rise of smartphones, there are still relatively low levels of financial literacy.

The banking sector also faces headwinds globally. The 2019 report on banking by McKinsey & Company indicates that over half of the world's banks are in a weak position even before any projected economic downturn. This is amid a weakening global economic climate. In April 2020, the IMF forecast global economic growth to be -3%, significantly worse than that experienced during the global financial crisis of 2008–9, constituting the worst recession since the Great Depression, and in stark contrast to the 3.6% expansion recorded in 2019. In fact, according to a Bloomberg article on the McKinsey & Company report, 'a majority of banks globally may not be economically viable because their returns on equity are not keeping pace with costs. It urged firms to take steps such as developing technology, farming out operations and bulking up through mergers ahead of a potential economic slowdown'.[4] Of course, all of this has been worsened by the impact of the coronavirus globally.

McKinsey & Company warned that banks risk becoming footnotes to history as new entrants change consumer behaviour and most recent attempts by banks to boost efficiency appear to have been little more than 'business as usual'.[5] To put this into numbers, banks allocate only 35% of their IT budgets to innovation, while fintech companies spend more than 70%. The grim outlook has already spooked investors, the report showing that globally, banks' valuations have fallen between 15% and 20% since the start of 2019. Banks must adapt, and, in order to do so, they need to adopt technologies such as AI.

AI and banking

According to a report by Autonomous Research, by 2030 AI technologies will allow banks to reduce operating costs by 22%. This is expected to save financial institutions around US$1 trillion in the long run. The process of client communication, for example, is driven by language-processing software that is programmed to determine a client's needs. Fintechs utilise AI to generate revenue, to refine products and services, and to develop new business value chains. A downside of the utilisation of AI could, however, be a reduction of reliance on human beings. Although the concept of a robot adviser for clients is one that is still in the making, using AI capabilities, the profile of a customer or client can be analysed to provide customised advice on major financial decisions. Simple calculators on banking websites, for instance, are applications that have enabled ordinary consumers to calculate the affordability of property, to determine whether they qualify for loans, and to have access to information about interest rates and repayment terms. Such calculators are useful in that they save banks time and cut down on the need to deploy staff to provide consumers with simple calculations, demonstrating how the utilisation of algorithms is embedded in most fintech applications. Major financial institutions, of course, use more sophisticated varieties of algorithms for decision-making. The use of chatbots by banks allows just-in-time queries – information provided exactly when it is needed – to be managed efficiently. Call centres, for example, rely on bots to listen to calls and offer possible answers using neural networks. At the back end of a bank, there are machine-learning algorithms that can detect fraudulent transactions or behaviour that may have otherwise gone undetected. Banks, therefore, already use AI technology to detect fraud and to conduct forensic audits.

The banking system works by accepting deposits from clients, borrowing money from the Reserve Bank, and raising money from the capital markets by issuing bonds and then creating credit to lend to whoever applies and qualifies for loans. For the bank to be profitable, the effective lending rate needs to be higher than the effective borrowing rate. For the bank to be efficient, it must employ competent individuals who understand credit and debit, interest rates, repo rates and the banking system, for instance, so that they can create products that can be sold to clients. The advent of AI is now reducing the need for people in the banking system. Recently, Michael Jordaan and Yatin Narsai launched Bank Zero, which has no branches and operates only in the digital space. According to the South African Banking Sentiment Index 2019, published by BrandsEye, South Africans are becoming increasingly cynical about traditional banks. Players, such as Bank Zero, are hoping to fill this gap. As BrandsEye CEO Nic Ray put it, 'The excitement that paved the way for their long-awaited arrival will likely be tempered by growing service requests and complaints as the banks onboard new clients. The key challenge will be maintaining a positive client experience as they scale their client base and switch on new products and services.'[6]

In South Africa, new entrants, such as Bank Zero, as well as Discovery Bank and TymeBank, are leading the charge in this field and will determine whether this works for South African consumers. According to BrandsEye, the response to Discovery Bank's launch was underwhelming. In contrast, TymeBank performed well, with growing awareness and improving sentiment post-launch, with further positive sentiment for Bank Zero driven by engagement with Michael Jordaan. According to research from S&P Global Rating, increased digitisation and greater efficiency are expected to support the performance of South African banks.

To stay ahead of the game and meet clients' needs, banks simply cannot afford to pay for costly and largely underused branches. Instead, the focus needs to shift to bettering their online offerings. According to Business Insider Intelligence's *AI in Banking and Payments* research report, the aggregate potential cost savings for banks from AI applications is estimated at US$447 billion by 2023, with the front and middle office accounting for US$416 billion of that total. An essential purpose of banks is to use data – such as information relating to transactions, purchase history, customers' social media and mobile-banking usage – to provide better client services, detect fraud and understand consumer sentiment.

For example, for a bank to decide whether or not to offer a customer credit, it needs to evaluate the creditworthiness of the individual. This is done so that the bank minimises the chances that the client will default. Conventionally, human experts estimate the income of the individual, his or her monthly expenses, and look at other factors, such as where he or she lives, for example, and then gives him or her a credit score. This credit score determines the interest rate this person will pay on the loan. In my book, *Economic Modeling Using Artificial Intelligence Methods,* I describe multi-agent AI systems that have been able to perform credit scoring better than humans.[7] This is because they are able to internet-search information on the prospective client and incorporate this into the individual's credit score.

Banks have to make their decisions with incomplete/imperfect information, and with clients about whom they have limited knowledge. This phenomenon is known as 'information asymmetry'. Their work on this concept won Joseph Stiglitz, George Akerlof and Michael Spence a Nobel Prize in Economics. AI breaks down the barriers of information asymmetry and promotes fair trade in the banking sector. Behavioural economics has taught us that humans can never be entirely rational and that they make their decisions based on truncated logic and information. This idea of the limited rationality of humans won Daniel Kahneman and Herbert Simon Nobel Prizes in Economics. The influence of human behaviour in the market is to curtail the efficiency of those markets. The replacement of human decision-makers with AI decision-makers ensures that markets will be more efficient and thereby promotes fair trade. If markets are not efficient, people can gain from the market without putting in the effort, and this can result in inefficient allocation of resources and thereby curtail productivity.

Banks sell financial instruments called derivatives, and one example of these is options. An option is a contract that gives an entity a right and not an obligation to buy or sell a particular good, at a specific time and at a particular price (called the strike price). Pricing these options has been a difficult task, and Myron Scholes and Robert Merton were awarded the Nobel Prize for their work on this. But the advent of AI has made it possible to price these instruments better; the emergence of fast computers has made it possible to simulate the behaviour of these instruments. Recently, UJ developed a mechanism of pricing options with imprecisely defined parameters using the type of AI known as fuzzy logic.

The digital bank

It is hardly surprising that there is a push for a more digitised offering from banks. The banking sector, arguably, has seen the most significant overhaul with the 4IR. This, of course, is the sector from where much of the immense fear has stemmed. As branches become somewhat obsolete, there are increasing worries around job losses. Threats of a banking strike in South Africa in 2019 all but confirmed this, after several retail banks announced retrenchments. It is hardly surprising when you consider the demand of consumers.

Stock exchanges and related services in the financial sector use AI seamlessly to manage sophisticated trading decisions. Given the swift nature of the traditional trading floor, more than 99% of trading decisions are based on access to significant datasets that provide information swiftly to a trader.

In Japan, the Nomura Securities Company uses AI robot traders for high-volume trading.

The deployment of AI in financial services enables clients to have immediate access to valuable information. The ability of AI to detect patterns in data has the potential to increase accuracy, which is critical to this sector. While fintech companies are rapidly adopting AI, there is a shortage of human resources with the ability to understand both the financial services industry and how to utilise technology. It requires a variety of skills, which then begs the question as to whether universities are producing graduates who are able to embrace AI, the world of business and the financial services industry. In 2019, the South African Institute of Chartered Accountants (SAICA) and UJ launched a series of short learning programmes to equip accountants with skills to navigate the 4IR. Titled '4IR for Accountants', topics include AI, machine learning, natural-language processing, blockchain and ethics. It is worth noting that some would argue that such programmes should have been conceptualised earlier in the game – after all, the financial services sector has long embraced the 4IR.

In accounting, for example, there is a move towards using machines that analyse balance sheets and detect fraud. AI presents an opportunity to streamline this process through automating data entry and reconciliations, for instance. Undoubtedly, this will disrupt the industry in a significant way for the first time since the advent of bookkeeping. Machines are picking up the time-consuming, repetitive aspect of accounting, and this presents an opportunity to eliminate accounting errors, thereby ultimately reducing the liability of accounting firms and allowing them to take on more clients. The use of AI could give rise to a whole new class of products and services explicitly applicable in the areas of accounting, such as customer service, research and development, logistics, sales, marketing and information analysis. This also fundamentally changes the role of accountants. Instead of simply performing the traditional role of crunching numbers, accountants could move into more of an advisory role, offering value-added services to clients.

Rethinking payment

In 2019, the Gautrain Management Agency announced that Gautrain commuters would be able to use the tap-and-go function on debit and credit cards to ride the train. There would be no need to buy a Gautrain card – which requires a minimum amount of R27 and expires every five years even if you are still using it – or to stand in a queue to top it up. As a result of this, there is now no need to worry about whether or not you have loaded enough money onto your card for a roundtrip and parking. Instead, much like the underground system in London, you merely touch in at the start of your trip, and touch out at the end of it, on the card readers.

Adopting this kind of technology is vital if we are to stay relevant and compete with the likes of the UK and the US. In the London Underground, commuters are able to pay for their fares using contactless cards. They can also make mobile payments using devices such as smartphones or smartwatches. Apple Pay has emerged as the most popular payment app in the US, with an estimated 30.3 million Apple-device owners using the tap-to-pay option in the country.[8] This is even more remarkable when you consider that, in 2019, Apple Pay surpassed the Starbucks payment app as the most popular mobile payment method for shoppers at US stores.[9] Starbucks developed their app in order to make it easier for customers of the popular brand to pay for goods, but obviously it can only be used at Starbucks – unlike Apple Pay, which can be used at a range of retail outlets. But, while Apple Pay leads, it is not the only option. There is also Fitbit Pay, Garmin Pay, Google Pay and Samsung Pay, for example.

Although Apple Pay has not yet been launched in South Africa, Samsung Pay, which was launched locally in 2019, has gained some excellent traction. As of March 2019, it had already generated 400 000 purchase transactions since launch.[10]

Africa's digital economy

Africa's Pulse, an analysis conducted by the World Bank, suggests that digital transformation could increase regional economic growth by up to 2% per annum, providing fresh opportunities for inclusive growth through innovation, service delivery and job creation.[11] There are already vast technological hubs and start-ups driven by youth across the continent. Take Stellenbosch, for example, which some have labelled the Silicon Valley of Africa. Here, successful fintech start-ups such as Entersekt, which pioneered mobile-based authentication for payments, or Snapscan, which allows you to make payments at restaurants or retailers on your phone by scanning a QR code, have emerged. Still, there is room for more and I would be remiss if I did not allude to the challenges we face on the continent. Despite good mobile connectivity, access to data and the emergence of tech hubs, Africa is still lagging behind other digital economies around the world, with comparatively low access to digital technologies.

For example, *Africa's Pulse* shows that eighteen of the twenty least wireless-connected countries in the world are in Africa. In countries with high levels of wireless connectivity, 90% of homes have access to fixed-broadband services. In contrast, in African countries, only 2% of households have such access. This figure drops to below 1% in 16 out of the 55 countries in Africa. Fixed-broadband subscriptions, essential for more extensive data needs of growing businesses, sit at only 0.6%, while Africa's level of internet bandwidth used represents just 1% of the world's total.

Added to this, the WEF's *Readiness for the Future of Production Report 2018*, which analysed countries' readiness to exploit emerging technologies, showed that out of '25 African countries assessed, 22 were classified as having a low level of readiness for the future, due to lack of the necessary enabling conditions'.[12] While in many instances, Africa is leading the charge, the numbers indicate that there is a fundamental need for further investment in the digital economy for us to keep pace with the rest of the world.

There are already some solutions, however. For example, digital businesses could target customers outside of Africa with a localised offering. Andela, which was started in Nigeria, for instance, identifies and supports software developers in Africa, who are then outsourced to tech companies around the world. While in 2019 Andela announced that 250 junior developers would be laid off in Nigeria and Uganda, and approximately 170 could be affected in Kenya, it has been a significant job-creator for senior developers. A few months later, the company expanded into Egypt and launched its first remote working centre on the continent. Engineers on a four-year contract are absorbed into the team and go on to provide their skills to more than 200 international clients, which have included Viacom, Cloudflare, GitHub and InVision.[13]

Elsewhere, there is scope for technology that can be adapted to a local context. M-Pesa – launched in countries such as the Democratic Republic of the Congo, Egypt, Ghana, Kenya, Lesotho, Mozambique and Tanzania

– enables clients to send, receive and store money safely and securely via a basic smartphone. Users can even top up airtime, pay bills, receive wages and salaries or get a short-term loan.

Another solution is to make data more accessible to all segments of the population. In October 2019, Mobiz CEO Greg Chen argued in an article for Tech Financials, '[One] issue that, if addressed, has the potential to bring about real and lasting change is data inequality. Put simply; mobile data is relatively affordable for wealthy and middle-class South Africans and prohibitively expensive for the country's poorer citizens. Given this, large swathes of the population are being denied access to tools and services that could empower them to find and create jobs'.[14]

Tackling financial inclusion

It is clear then that digital banks in South Africa need to address financial inclusion. Around 90% of people in South Africa have access to financial services, and 80% have bank accounts, according to data by FinMark Trust in 2018. However, many of these accounts are not active, and 63% of people make use of unregistered moneylenders known as 'mashonisa' (loan sharks). The use of technology in the sector could curb this. For example, over 275 000 people now use uKheshe, a micropayment platform established in 2018. According to trade body GSMA, three-quarters of the population in sub-Saharan Africa have a SIM connection, while around a third of mobile users have a smartphone. The number of mobile internet subscribers in the region has quadrupled since the start of the decade.[15]

CHAPTER 15

Taxation

In February 2020, SARS announced that it would be embarking on a programme and recruitment drive to incorporate data-driven insights, machine learning and AI into its fight against tax evasion. This came just months after it introduced an app for filing tax returns in a bid to reduce traffic at its branches. The app allows taxpayers to, among other things, register for e-filing (electronic filing), submit a tax return, upload supporting documents, make payments to SARS, view the status of returns, and access a tax calculator. As the commissioner of the tax agency, Edward Kieswetter, put it at the time, 'We cannot ignore the power of a data and technology-enabled organisation, and the impact it will have on the future world of work. We can, however, prepare for it by consciously and actively managing the interplay between human effort and artificial intelligence.'[1] This has been part of the revenue service's shift towards a more significant digital offering as it rebuilds its reputation following years of mismanagement. Yet, because of the weak economy and governance failures in South Africa, revenue collection has fallen short of budget targets since 2014. SARS was once regarded as a world-class institution and had a reputation of overshooting targets, releasing resources for tax cuts and capital expenditure. This development is hardly surprising given that South Africa's economy has not grown by more than 2% since 2013 and that former commissioner Tom Moyane was fired in November 2018, in line with recommendations from Judge Robert Nugent, who headed a commission of inquiry into governance failures at the institution.

While this is a new strategy in South Africa, the move towards integrating AI into tax systems has already been adopted by tax agencies around the globe. Tax agencies deal with a great deal of data that no longer encompasses only traditional sources, such as personal tax or corporate tax (taxes paid by companies), but now also includes digital payments, electronic invoicing, online cash registers and point-of-sale solutions, for instance. As Kieswetter put it, 'We have enormous islands of data that we can impose AI on to help us enforce compliance.'[2]

Tax collection is the most significant way in which the government generates its revenue. Before the final audit, in South Africa, the tax collected in the year 2016/2017 was R1.216 trillion. This tax is used to run the state, pay civil servants, run public schools and hospitals, maintain roads, as well as pay social grants. These are public goods because they are our investment in the stability and prosperity of our nation. If tax collection is higher than our expenditure, then we run a budget surplus, and, of course, if our tax collection is lower than our expenditure, we run a budget deficit. A persisting deficit in a country ultimately renders the government bankrupt. But revenue collection is based on macroeconomic conditions, such as growth figures and the government's debt, tax proposals and taxpayer compliance. The shift to AI makes it possible for organisations such as SARS to collect taxes from citizens in a shorter time and reduce the number of tax defaulters, increasing the revenue service's effectiveness and efficiency.

The impact of the 4IR on tax collection

As we have already seen, it has been argued that the 4IR will reduce employment and thus potentially reduce the amount of collectable tax. Any task that typically does not involve more than one minute of thinking to complete will be automated and will be performed by a machine. This era of the 4IR will be a post-work one, because the need for humans in the workforce will be severely curtailed, effectively changing the face of labour. AI robots will populate factories and, unlike human workers, will not be organised by trade unions. To paraphrase Karl Marx, AI robots cannot be mobilised by the cry: 'AI robots of the world unite! You have nothing to lose but your chains.' Our participation in this revolution will not be optional. Either we participate, or as a country we are economically obliterated, relegated to the 'dustbin of history'. Those who adopt the means of production of the 4IR will produce goods at such low costs that the price of products will fall sharply. Those who refuse to use AI robots to run their factories will be so uncompetitive in price that no one will buy their goods, and they will ultimately be put out of business by market forces.

The social consequences of the 4IR will be extensive. Those with financial capital will simply buy AI robots and produce goods and services to maximise profit. So, the concept of the poor getting poorer and the rich getting richer will thus be exacerbated by the unconstrained drive towards the 4IR. The Gini coefficient – which is a measure of inequality in society – will increase, threatening the very existence of the notion of a nation state. Now, if people are going to lose jobs, who will buy the goods that these robots will produce? Where will customers get the money to buy these goods if employment, as we know it, no longer exists? Leading capitalists have already proposed a solution. Bill Gates, founder of the giant computer company Microsoft, proposes the taxation of AI robots – discussed later in the chapter – while South African-born technology billionaire Elon Musk proposes introducing the universal basic income (UBI). But will these measures be enough? Perhaps the first port of call is to understand the impact of these revolutionary changes on taxation.

For example, neither taxing the use of robots nor imposing taxes on automation that results in human labour being replaced will result in tax revenues declining. The loss of jobs, for example, could result in widening income inequalities, and not imposing penalties for automation or robots could result in a myopic move on the part of policymakers in terms of expanding the tax basket. However, an argument could equally be made for avoiding knee-jerk responses, such as imposing a 'robot tax', because that could, in turn, create a perverse downward trend in terms of productivity and a long-term hindrance to the progress of a country. So, the taxation argument does not really hold as a quick fix to automation replacing workers. What is definitely required is reimagining our education system to offer continuous training and reskilling. It requires educational institutions to partner closely with industry to identify future skills and even jointly imagine future jobs. Displacement of staff by automation can be seen as a window of opportunity to provide new skills and retrain for new functions. The current fear of job losses prompted by disruptions of the kind predicted in the 4IR is not new. History shows that each society has continuously had to be agile and change with the times and developments. Now more than ever, there is a need for a country to wisely invest in technological revolutions as well as its citizens.

Imposing taxes on automation in order to optimise tax revenues

In 1974, American economist Arthur Laffer proposed a theory showing the relationship between the effective tax rate and the tax revenue collected by government – although, it has to be said, the ideas Laffer popularised were not new and were described as early as the 14th century by Arab scholar Ibn Khaldun. The Laffer curve states that if the government has a tax rate of 0%, it will collect no tax at all. If it imposes a tax rate of 100%, people will have no incentive to work, and the taxes collected will also be zero. Somewhere between the effective tax rate of 0% and 100%, there is a tax level that results in the maximum tax collected. Without automation, this maximum tax collected will be a certain number. If we automate our factories without changing the corporate tax rate, then the amount of tax collected will decrease. If we automate our factories and responsibly increase the corporate tax rate, we can end up with the amount of tax collected being higher than that obtained without automation.

Automating tax collection and practices

Many of the repetitive aspects of tax collection and practices can be eliminated by automating them. We have already seen accountants adapt to change brought about by technological innovations, such as the online tax-filing system that SARS has pushed for in order to automate its processes. As we have seen, the move towards e-filing in recent years has not put accountants out of business but instead streamlined the process, making it possible to file many more returns than they could before. This essentially reimagines the role of the accountant. According to a 2014 CPA of the future study, accountants were already preparing for 'unprecedented, massive and highly accelerated change'.[3] Some 80% of the participants in the study agreed that their roles will change substantially over the next decade, with more emphasis on consultative business development, risk management and advisory functions. It also found that around 80% of CPA firms were concerned about recruitment to meet future needs.[4]

The big question thus remains: how will intelligent automation in industries impact tax collection? Suppose an auto-manufacturing plant in the City of Tshwane has 100 people, who each contribute R5 000 per month in taxes to SARS. If this plant decides to automate the jobs of these 100 employees, then the tax revenue for SARS will drop by R500 000. The only way this tax revenue will not drop is if SARS correspondingly increases corporate tax for this company by R500 000. The concept of increasing corporate tax because of automation means that we are effectively taxing AI robots. As Bank of England economist Andy Haldane put it, 'The rise of automation is a regressive income tax on the unskilled.' In an interview with *Quartz* in February 2017, Bill Gates argued in favour of a robot tax, saying that governments should tax companies' usage of robots as a way to slow down the spread of automation temporarily and to fund other types of employment.[5] As Gates put it, 'Right now, the human worker who does, say, $50,000 worth of work in a factory, that income is taxed, and you get income tax, social security tax, all those things. If a robot comes in to do the same thing, you'd think that we'd tax the robot at a similar level.'[6] Gates's stance came as a surprise to many, given that Microsoft is a leading player in AI technology. In fact, in the same year, EU legislators reflected on a suggestion to tax the owners of robots in order to finance the training of workers who have lost their jobs as a result of automation. Ultimately, however, this proposal was rejected – a decision hailed by the robotics industry, which insists that it would stunt innovation. South Korea limited tax incentives for businesses investing in automation, which many compared to a robot tax. This is not, however, a robot tax in the traditional sense in that it does not directly tax companies. However, the 2% drop in the available tax deduction to these companies was announced as a way to slow down the growth and pace of automation. Yet, much of the automation of jobs originally meant for humans largely happens as a slow-creep process that is invisible to most people in society, since it happens so gradually. For example, major cities once employed meter maids – so, naturally, taxpayers – to monitor parking bays, but over the years, the introduction of parkades and the reduction in the need for supervised parking meters has led to the disappearance of meter maids. This involved the evolution of several different technologies over time, and so it would be impossible to recover an automation tax in this instance at all. While there are arguments for and against 'robot tax', there is a stark reality accentuated by our newspaper headlines regularly in South Africa of high unemployment and job losses, which also points to a shrinking economy.

Automating the economy

Can we realistically automate an entire economy? To answer this question, we should study the work of Nobel laureate Arthur Lewis, who hailed from St Lucia in the Caribbean, and who became the first person of African descent to win the Nobel Prize in Economics. Lewis proposed what later became known as the 'two-sector economy theory'. In this theory, he studies how to move an economy from agriculture to industry, and he suggests that by moving labour from the agricultural sector to the industrial sector and using the resulting profits to expand industrial production, a country can move from a developing to a developed economy. He proposed that this could be done until it is no longer economically feasible to move labour from an agricultural to an industrial sector – the point now referred to as the 'Lewis turning point'. After the Lewis turning point has been reached, the cost of labour in that economy starts rising. China used this theory for its developmental agenda and reached its Lewis turning point in 2009, after which the cost of labour started rising. In our book *Artificial Intelligence and Economic Theory*, Evan Hurwitz and I apply Lewis's theory to automation and define its limits.[7] We divide the economy into two parts: the AI-machine economy and the human economy. Labour will move from the human economy to the AI-machine economy until it no longer makes economic sense to migrate labour from humans to AI machines, possibly because the tasks involved are too complicated or too human to be automated. A good example is the patient-doctor scenario; for many of us, the patient-doctor relationship is a human one and not one that we would readily give up in favour of robot doctors. This means that the limit of automation is a point at which it is more expensive to automate a job than to use humans, mainly due to the difficulty of deploying the technology.

Given the above, what does the South African government do to ensure that we collect enough taxes that will provide for a growing economy and, consequently, a stable society? Firstly, we need to have dynamic and world-class economics scientists who are sufficiently skilled in modelling economic, social and political phenomena. Secondly, we need to train enough people who understand automation and AI. Thirdly, we should use AI robots to detect and prevent tax evasion. Fourthly, we should develop a framework for taxing companies that are domiciled overseas but make their money in South Africa. These include international companies operating in South Africa, such as ride-hailing service Uber, and video-streaming service Netflix, also based in California. The regulatory frameworks of countries have not sufficiently caught up with new business models for the digital economy. For example, in 2019, a conservative estimate of unpaid VAT in the UK was pegged at over £1 billion, just under R23.5 billion. In fact, Uber's entire business model is predicated on keeping prices low by avoiding tax payments. This impact on our tax regime has not been tested in South Africa as yet, and, despite the occasional flare-up by the public, regulatory tax bodies have not moved to address this. Tax not generated is income lost to the country that could be used for the public good of the citizens.

It can be argued that, given the limited tax base of South Africa, there is a need to use AI to combat tax evasion, fraud and non-payment. It was estimated in 2019 that there were just over 22 million taxpayers registered with SARS, yet, at the end of the year, just over 4.9 million had submitted their returns. To this end, AI can be used successfully for predictive modelling to identify potential fraudulent evasions and for fraud detection. This would require working on big data, which would include accessing historical data on tax returns, information about financial transactions across multiple sets of data, and various other databases that could be scanned for patterns, company data or individual profile information. While these forms of machine learning do not replace the role of the tax auditor, they can contribute to greater efficiency. For example, tax evasion is very much a cat-and-mouse game. With each introduction of new tax rules, there is often significant manoeuvring, especially by large corporations, in order to avoid payment. The work of auditors and tax officials would be greatly enhanced by the use of different forms of machine learning to detect and trace such evasions. The work of officials could further be bolstered by the sophisticated usage of data mining and analytics, both of which techniques are rooted in statistics and computer science. Cloud computing makes analysing big data easier. Deloitte, for instance, has developed 'Revatic Smart technology', a tool that tax-collection agencies can use to automate the process of checking companies' compliance in submitting VAT returns. Through a complex automated process that saves these agencies time and money, the technology scans receipts and invoices in order to determine correct VAT inputs.

These are just a smattering of some of the techniques that could be used by countries' tax authorities. The solutions offered by the 4IR range from simple digital interactions with customers to process refinements for taxpayers filing returns or corporations submitting tax returns.

Again, it is clear that we need to deploy technology more efficiently in tax collection, while acknowledging that this will require the bolstering of the infrastructure of, for example, SARS. Despite the automation of some front-office services and interactions, greater efficiencies can be achieved by optimising the usage of AI. The assumption is, of course, that alongside the deployment of technology will be the need to ensure that the workforce is also sufficiently trained. An argument pervading this chapter is that a country needs a holistic assessment of the direction in which it is going and a more precise identification of how the maximisation of the tax sources can be enhanced.

CHAPTER 16

Market efficiency

Towards the end of 2018, it emerged that then Minister of Finance Nhlanhla Nene had asked President Cyril Ramaphosa to be released from his position. It was what some might term 'a perfect catastrophe' for Nene. His gripping testimony at the state-capture enquiry revealed that, while he had stood up to former President Jacob Zuma and called him dishonourable, he had also met with the now-infamous Gupta family at their Saxonwold home, despite previously denying this. To many outsiders, this would seem inconspicuous, but in the fraught world of South African politics, Nene could not claw his way back from this damning revelation. The Guptas have been accused of working with Jacob Zuma to secure billions of rand in government contracts and determine cabinet appointments. While there was no suggestion that Nene had done anything illegal, the association with the controversial family was enough to crucify him. With the daggers out for him, he opted to fall on his sword. To exacerbate the matter for Nene, following his testimony to the state-capture commission, the *Mail & Guardian* newspaper and the amaBhungane Centre for Investigative Journalism reported that his son, Siyabonga Nene, had requested the Public Investment Corporation (PIC) to finance a deal to acquire a stake in an oil refinery in Mozambique. As President Ramaphosa accepted the resignation 'in the interests of good governance', the JSE All-Share Index declined to its lowest levels in 2018, while the rand also weakened.

In October 2018, when President Ramaphosa announced that the former governor of the Reserve Bank, Tito Mboweni, would replace Nene as Minister of Finance, the rand gained 30c against the euro, and 20c against the dollar and the British pound. This was surprising, given Mboweni's seeming reluctance; just a few months before he had tweeted, 'Against the wisdom of my team, please don't tell them this. It's between us, I am not available for Minister of Finance. You cannot recycle the same people all over again. It is time for young people. We are available for advisory roles. Not cabinet. We have done that.'

Perhaps it is only if you are familiar with South African politics that you might see what a roller-coaster this has been. Mboweni was appointed as South Africa's fifth finance minister in less than three years. The ride to that was not smooth. It all began in December 2015 with what is now referred to as 'Nenegate', when Nene was unceremoniously axed by Jacob Zuma and replaced by little-known MP Des van Rooyen, who infamously became known as 'Mr Weekend Special'. That move saw the rand crash, as investors pulled billions out of the country's bond and stock markets. It was political analyst Daniel Silke who said, 'It was a jolt to the collective nation and its shockwaves of political interference, crony capitalism, rent-seeking and poor governance were in sharp focus.'[1] This was repeated in 2017 when then Finance Minister and current Minister of Public Enterprises Pravin Gordhan was fired in a surprise midnight cabinet reshuffle. As journalist Chris Bateman put it, 'South Africa seems to change Finance Ministers like it does shirts, some of them soiled, others clean and in the latter years, a Zuptoid camo-shirt – hard to spot the dirty patches.'[2]

Markets behave strangely. They respond to what is happening and very often in an irrational way. Are markets rational? Are they efficient? What do rational and efficient markets mean?

Rational markets

The 19th-century English philosophers Jeremy Bentham and John Stuart Mill proposed the utilitarian theory, which is used to define rationality. According to this theory, an agent that is rational maximises utility or, in other words, maximises the usefulness of a particular good or service. For example, if you have R100, a rational agent will purchase the combination of goods that will provide them with the most satisfaction. Similarly, a rational market maximises the allocation and distribution of resources. Economics Nobel Prize winner Eugene Fama proposed the 'efficient-market hypothesis', which states that a perfectly rational market is efficient. In order to understand this theory, it is important to understand the converse. If markets are not efficient, resources may be misallocated. This results in deserving traders, or those who use the markets to the benefit of the economy, losing out while underserving traders, or those who use the markets to the detriment of the economy, gain. This ultimately results in negative economic growth.[3]

Simply put, investors should earn a return on their investments according to their perceived risk at the time of investment. The challenge then is for investors to be privy to the same information – which is why publicly listed companies have to declare everything publicly. The JSE, for instance, uses the Stock Exchange News Service (SENS) to provide users with access to company announcements, such as mergers, takeovers, rights offers, capital issues and cautionary – all of which have a direct impact on the movement in the market.[4]

Why do inefficient markets harm the economy? In the ten months preceding the end of 1999, British company Lastminute.com, which specialised in last-minute online travel and leisure bookings, such as for flights and restaurants, was generating £330 000 in income. Lastminute.com was then floated on the London Stock Exchange, and its valuation reached £768 million before it crashed when the dot-com bubble burst following excessive speculation in internet-related companies in the late 1990s, based on the massive growth in the use and adoption of the internet. Yet, many of these stocks were overvalued and there was little focus from the market on how these companies generated revenue. This saw the biggest-ever one-day fall in the New York Stock Exchange at the time, and Lastminute.com ultimately traded at 30% of its flotation price. The market was irrational in its view of this company and, consequently, inefficiently allocated resources to it, a development that was bad for the economy. This market inefficiency was due to inaccurate information on the value of the Lastminute.com company.

The Steinhoff International scandal is another example of market inefficiency caused by limited information. The company dropped from a market capitalisation of R104 billion in 2015 to R9 billion in 2018; the management of Steinhoff International apparently deliberately withholding information from the shareholders. As a PwC investigation found, 'A small group of Steinhoff Group former executives and other non-Steinhoff executives, led by a senior management executive, structured and implemented various transactions over a number of years which had the result of substantially inflating the profit and asset values of the Steinhoff Group over an extended period ... It appears that the Steinhoff Group entered into several transactions (some of which were fictitious or irregular) with allegedly independent third party entities which resulted in the inflation of profits and asset values.' This proved costly for the South African economy. For instance, the government's total exposure in the Steinhoff saga was estimated at around R26 billion in December 2017. In addition, there was a dramatic knock-on effect, with massive losses for pensioners, investors and taxpayers – hardly surprising when you consider that before its collapse, Steinhoff was the seventh-biggest company in the FTSE-JSE Shareholder Weighted Index (SWIX), a benchmark used by professional investors.[5]

The efficient-market hypothesis is not able to adequately explain economic bubbles. Take the global financial crisis of 2008–9, for instance. In 2007, the US saw a significant depreciation in the subprime mortgage market, which then spiralled into a full-blown international banking crisis. This led to the collapse, on 15 September 2008, of the investment bank Lehman Brothers, which contributed to the loss of US$10 trillion in market capitalisation in a single month. This triggered a global economic downturn. In the lead-up to this, lax credit requirements spurred a housing boom that drove speculation, pushing up housing prices and creating a real-estate bubble. A bubble such as this occurs when the price of assets rises higher than its actual economic value and then falls rapidly. In contrast to the efficient-market hypothesis, in a bubble, the right price is fundamentally different from the market price – in other words, the consensus price is wrong. As Adair Turner, then chairman of the Financial Services Authority (FSA), told the *Financial Times*, 'Financial instability is driven by human myopia and imperfect rationality as well as by poor incentives and because any financial system will mutate to create new risks in the face of any finite and permanent set of rules.'[6]

But, again, is the South African market efficient? It is generally believed that markets are not efficient. Warren Buffett famously said, 'I'd be a bum on the street with a tin cup if the markets were always efficient.' As he put it to his shareholders in a letter in 1998, '… naturally, the disservice done to students and gullible investment professionals who have swallowed Efficient Market Hypothesis has been an extraordinary service to us. In any sort of a contest – financial, mental, or physical – it's an enormous advantage to have opponents who have been taught that it's useless to even try. From a selfish point of view, [we] should probably endow chairs to ensure the perpetual teaching of Efficient Market Hypothesis.'[7] There are several reasons why this is the case, and one of these is that markets are influenced by the often-erratic behaviour of people who participate in them. In essence, markets reflect real information, such as the output of the mining sector, which is rational, and imaginary factors, such as the attitude of traders towards mining, which can be irrational. It also assumes that all investors perceive all the available information in precisely the same manner. The mathematical field of 'complex analysis' prescribes that nature can be represented in two forms: the imaginary and real axes. The behaviour of the market, going up and down because of the changes in cabinet ministers, for example, is because the market is also driven by imaginary factors emanating from human behaviour. For instance, would there have been a different market reaction had someone else been appointed finance minister instead of Mboweni? Some of the names touted at the time were Barbara Creecy, Daniel Mminele and Mcebisi Jonas. Would these individuals have thrown a spanner in the works?

In his book *Thinking, Fast and Slow*, Daniel Kahneman studies the behaviour of human beings when they make decisions.[8] Some of the observations he makes are that humans truncate complex problems, solve simple ones and, when they make decisions, focus on avoiding losses at the expense of making gains. This simply means that human beings do not make rational decisions because they never maximise utility. This also means that human behaviour significantly influences the markets, which are operated by humans and cannot, therefore, be efficient.

Markets in the 4IR

The JSE used to be located in downtown Johannesburg, and trades here were made by human beings shouting for particular stocks at particular prices. In order to be successful at that stock exchange, you needed to be tall and have a loud voice – in fact, recruiters specifically looked for height and voice, which was of course discriminatory. The JSE then moved to Sandton and entered the electronic age. With this shift, height and voice were no longer competitive advantages; instead, digital literacy, data analytics and AI became competitive advantages. For this new JSE, trading in close proximity to Sandton became a competitive advantage, because if someone in Thohoyandou trades via the internet, then his/her trade arrives at the JSE slower than the trade made from Sandton. Today, people are making these trades at a faster pace, resulting in high-frequency trading. In my book *Economic Modeling Using Artificial Intelligence Methods*, I describe how AI is changing the practice of trading and ushering in intelligent high-frequency trading. The effect of the confluence of faster computers, AI, high-frequency trading and advanced data analytics is that markets are more efficient than when only humans, and not machines, were involved in trading. Markets that use AI and advanced data analytics are markets of the 4IR. According to a study by UK research firm Coalition, electronic trades account for almost 45% of revenues in cash equities trading. Many hedge funds are also using AI-powered analysis to get investment ideas and build portfolios.[9] For instance, Trading Technologies in Chicago uses AI to instantaneously identify sophisticated trading strategies from data from multiple markets and provides its clients with an ongoing assessment of compliance risk.[10] In Seattle, Kavout uses a 'K Score' that analyses big data sets and generates many AI models to produce a stock-ranking rating. It recommends daily top stocks using pattern-recognition technology and a price-forecasting engine.[11]

So why are 4IR markets becoming more efficient, and what are the implications of this change on the economy and society? The first reason markets are becoming more productive is that we remove behavioural characteristics from the markets when we insert automation, AI and advanced data analytics into trading. With machine learning, for instance, you can analyse multiple earnings, interpret data and provide more in-depth insight. The second reason is that we are now able to understand the character of financial data. For a long time, we reduced the analysis of business data to simple statistical concepts, such as averages and variances. But, with advances in data analytics, we can now break data into its various components, and thus make machines understand data much easier. With advances in deep learning, we are able to analyse data in its entirety rather than in truncated and sampled forms. The third reason is advances in computing power, which makes it possible for financial companies to acquire supercomputing capabilities and be able to implement intelligent algorithms on big data more efficiently. The fourth reason is that we now have more information and more technology to analyse both structured and unstructured data. AI is also a more cost-effective tool for investors because it has much lower operating costs, and firms are also now able to implement automated market surveillance. For instance, in the US, the Nasdaq is using AI to detect irregular and potentially malicious trading activity. This will ease much of the tedious work of the Nasdaq's market-surveillance team, which reviews around 750 000 cases and identifies unusual price movements, trading errors and any potential manipulation.[12]

The impact of a more efficient market on the economy would be profound. Following the global financial crisis, for instance, many countries used monetary policy to combat the recession and reduced interest rates. The efficient-market hypothesis argues for stable and predictable monetary policy. More efficient markets can also stoke higher economic growth. Financial markets are essential in that they direct the flow of savings and investment in the economy, thus facilitating the accumulation of capital and the production of goods and services. As Asli Demirgüç-Kunt and Ross Levine put it in *Financial Structure and Economic Growth*, 'In particular, researchers have provided additional findings on the finance-growth nexus and have offered a much bolder appraisal of the causal relationship; firm-level, industry-level, and cross-country studies all suggest that the level of financial development exerts a large, positive impact on economic growth.'[13]

Of course, the rate at which markets are becoming efficient varies, depending on the state of development of countries. In many developing nations, for instance, there are limited financial markets, instruments and financial institutions, which makes it more costly to raise capital and, in turn, often lowers the return on investments – or, in other words, does little to stimulate economic growth. In his book *AI Superpowers: China, Silicon Valley and the New World Order*, Kai-Fu Lee describes how the US and China are becoming the only superpowers of the 4IR.[14] In this regard, if we do not invest heavily in AI and big data, countries that do will exploit inefficiencies in our own market – and to our detriment. Policy expert Eleonore Pauwels studied the geopolitics of AI and observed that the amount of money invested by the US into AI start-ups between 2012 and 2016 was US$17.9 billion, followed by US$2.6 billion by China and US$800 million by Canada. Pauwels concluded that, given this asymmetry of investments, countries that lag behind are going to be cyber-colonised. Coming back to the concept of market efficiency, countries that do not invest in the technologies of the 4IR will have inefficient markets, resulting in inefficient allocation and distribution of resources as well as an impaired economy.

What kind of skills do we need to produce a workforce that creates efficient markets and, consequently, a robust economy? Firstly, we need to provide a cadre of people who understand AI and big data and how these technologies interface with the economy. Secondly, we should develop leadership courses that reskill the existing workforce in industry, government and society so that they can better manage the 4IR markets as well as craft the associated legislation and strategies.

CHAPTER 17

Trade

Wednesday, 16 June 1976 was a turning point for South Africa. On that day, there were massive protests in Soweto against making Afrikaans a medium of instruction at schools. One of the outcomes of what became known as the Soweto Uprising was that a lot of young people went into exile. One of those was Mzala Nxumalo, who became one of the leading intellectuals of the 1976 generation. Mzala died in 1991, at the tender age of 35, after he had just completed his PhD thesis at the University of Essex in England and was planning for a postdoctoral fellowship at Yale. At Yale, Mzala intended to write a biography of ANC President OR Tambo. He was a formidable and dedicated writer, and I was introduced to his work in London in 1989, when I was seventeen years old and attending the London International Youth Science Fortnight.

As a young black boy dreaming of how I should take the African continent from an underdeveloped to a developed state, I was inspired by Mzala's writings. Mzala was a leader who believed that those who do not know, read or write, should not lead. Mzala was a scientific socialist, skilled in Marxist tools of analysis. A revolutionary internationalist of note, one of his celebrated works was a paper titled 'Cooking the Rice Inside the Pot',[1] in which he postulated that the war against oppression be fought at the home front. He also emphasised the need for international solidarity in fighting oppression in South Africa.

Mzala was not a physicist and, therefore, was not well versed on Max Planck's quantum theory, which states that an object can occupy two states at the same time. Quantum theory is an antithesis of scientific determinism, which says that everything can be analysed and explained by the chain of causes and effects that led to its occurrence. Mzala was interested in the national question, which, in turn, is an antithesis of quantum thinking in which an individual can occupy multiple identities at the same time. So, in terms of quantum thinking, you can 'cook the same rice inside and outside the same pot at the same time'. This is what scientists call the dual nature of being – we now know, for example, that light is a dual entity, existing both as a particle and as a wave. Science is beautiful, and we should study it to liberate ourselves socially, economically and politically.

Why is 'cooking the same rice inside and outside the same pot at the same time' an important concept in our times? In 2019, the media reported that the embassies of the UK, Netherlands, Germany, US and Switzerland wrote a letter to the South African president to complain about corruption in the country. This was an unprecedented move; foreign countries usually only step in when governments commit human-rights violations or act against the law. The letter warned President Ramaphosa that his investment campaign 'could fail unless SA starts to take tangible action against the perpetrators of state capture, corruption and other serious crimes'.[2] On taking up the presidency in 2018, Ramaphosa announced a campaign to lure US$100 billion in new investment within five years. By 2019, there was US$16 billion already committed and many more projects in the pipeline. In the letter, the five nations also stated that they were concerned about the challenges of foreign investment, referring to the Mining Charter, BEE targets and IP rights. What is startling about this incident is that 75% of our FDI depends on these five countries – which means that our economic growth and ambition are being 'cooked outside the pot by outside cooks'. For example, we assemble vehicles for companies such as Mercedes-Benz here in South Africa, but the parts of these cars are manufactured outside South Africa. So, to use Mzala's allegory, 'we are serving the rice that was cooked outside the pot'.

As we move forward in our long march towards economic freedom, we ought to find ways – and the right economic policies – that will move us forward. Countries with a more significant claim of the tech sphere appear poised to play a more prominent role in the 4IR – and, in many respects, are leading it.

The overhaul of global trade

Part of the change that will need to accompany the 4IR is an overhaul of global trade. We are in a digital age, that much is a given, and the continent boasts the demographic profile to lead this age. Much like a microcosm of the 4IR, which requires a convergence of humans and technology, this requires a convergence of forces across the public and private sectors.

However, as we move into this age, countries that are not seen as technologically ready could well see a drop in FDI. Technological readiness is based on a range of factors, from spending on education, to government policies, and the macroeconomic climate. Bhaso Ndzendze, a research director at UJ's Centre for Africa–China Studies, and I looked at the data on the ten best-scoring and ten worst-scoring countries in the WEF's annual Global Competitiveness Index between 2009 and 2018 to see whether their movement in the technological-readiness indicator was linked to FDI.

Though poor and technologically deprived countries can rewrite their own stories and catch up (or even lead) technologically, accounts of late developers managing to do this are a rarity. China (in present terms), Singapore and South Korea in the late 20th century, as well as Japan in the early 20th century, stand out for a reason. It is worth noting that these nations also did not engineer their economic miracles alone. The 'flying-geese' theory is a metaphor of how FDI moved from North America to Japan, and then to the other Asian countries, leading to economic miracles in its wake. This theory explains how a country can develop rapidly by focusing on labour-intensive industries that then become the basis for exports to more developed countries as the level of quality increases, ultimately leading to exponential economic development. It is in this way that American companies poured money into Japan.

We found that among the top-tier scorers, six out of ten of these countries saw mutual growth in improvements in 4IR readiness and FDI. Similarly, five out of the ten lower-tier countries saw a positive correlation, as well. This is encouraging because it shows that moving up in technological readiness is linked to FDI, regardless of where each country is placed on the ranking. This is, therefore, a window of opportunity. We further noted that both groupings had an equal chance of having their technological-readiness score decline and receiving less FDI than the preceding year. This took place 21 times for both groups out of a possible 77. Overall, based on the generalised findings among the twenty countries, it appears as though there is no rigid persistence of a global Matthew Effect. The 'Matthew Effect', or the accumulated advantage, is a social phenomenon that suggests that those who have more of an advantage acquire more or, in other words, the rich get richer, and the poor get poorer. Essentially, investment in technology sectors could yield high FDI.

So, where does Africa fit into this equation? According to the WEF, all fifteen African countries assessed for production readiness fall into the 'nascent' category. According to Deloitte, when we compare Africa to the rest of the world, the adoption of the 4IR is low.[3] Nonetheless, this is increasingly being acknowledged as important by economic and political leaders, mainly because of the impact that smart technologies can make at a socio-economic level. The greatest challenges in Africa identified by Deloitte are digital skills, accessibility and connectivity.

Out of the fifteen African countries ranked, only South Africa fell within the top 50 countries for two subdivisions. Since the 1980s, South Africa's manufacturing share of GDP has decreased from 25% to around 13% today. Nevertheless, the country still has the most substantial Structure of Production on the continent. Across the Drivers of Production component, South Africa's performance is mixed. On one hand, the ability to innovate is one of South Africa's greatest strengths, because the country has a strong innovation culture and formal entrepreneurial activity supported by a sophisticated financial sector. On the other hand, human capital remains the most pressing challenge in preparing for the future of production, because there remains a shortage of engineers and scientists, as well as digital skills. South Africa needs to improve its institutional framework in order to respond to change, offer a stable policy environment and direct innovation effectively.

We have seen a shift in recent years towards investment in sub-Saharan Africa. According to the World Bank, between 2000 and 2013, FDI flows into the region increased six-fold to US$45 billion, particularly in the manufacturing sectors. This showed a significant shift from the previous emphasis on the continent's natural resources, which saw investment flow into the extractive sectors. According to Accenture, harnessing digital technologies can generate R5 trillion in value for South African industries over the next decade, particularly in agriculture, infrastructures, manufacturing and financial services.

According to the Deloitte report, overall in the South African manufacturing industry, the adoption of smart technologies that accelerate the 4IR remains at the foundation stage, with some sector differences. There is an appreciation for advanced analytics within the automation and automotive sectors, but manufacturers have not yet explored the real opportunities for advanced analytics. The adoption of Cloud solutions is primarily driven more by consumers than businesses, the main concerns being the fear of cybercrime and privacy issues. Advanced sensor technologies are, with some exceptions, such as in the automotive industry, also still at a foundation stage.[4]

There is, however, interest among manufacturers to take advantage of the potential for better monitoring, controlling and tracking. Deloitte reports that the use of robotics is mostly at an automated stage and not yet at a smart or advanced stage, with no widespread adoption of additive manufacturing – such as 3D printing, which builds materials through a layering process – in South Africa, even though the significance and the potential of this technology is increasingly acknowledged.[5]

Perhaps a key step will be for African economies to harness digital trade through the African Continental Free Trade Area (AfCFTA) that came into effect in 2019. The agreement brings together all 55 member states of the AU, with a combined population of 1.3 billion people, including a growing middle class, and a nominal GDP of approximately US$2.3 trillion. Already, the possibility of new markets presents new avenues for tech start-ups and e-businesses. The key, of course, is to continue to harness this potential.

Trade in South Africa

International trade in South Africa represents 59.4% of the country's GDP. According to the World Trade Organization (WTO), South Africa exported US$93.9 billion worth of products in 2018, importing US$113.6 billion of goods. The country saw a US$1.6-billion trade surplus for the same year – a 65% decrease from 2017. According to a United Nations Conference on Trade and Development (UNCTAD) report, FDI into South Africa remained at around US$5 billion in 2019, which showed a slight decline in inflows compared to the previous year. As the report put it, this is because South Africa saw FDI more than double with the shift in leadership. Between 2017 and 2019, this represents an increase of 446%, with investments mostly in mining, petroleum refinery, food processing, information and communications technologies, and the renewable energy sector.

Yet, despite this surge in investment, the economy is still weak; it has not grown by more than 2% since 2013, and it was dealt another blow at the start of 2020 by the combination of more load-shedding and the COVID-19 pandemic. Credit rating agency Moody's Investors Service expected the global spread of the coronavirus to result in simultaneous supply and demand shocks, for instance. The biggest knock would come from China, one of South Africa's most significant trade partners. According to a report by PwC, because of its relationship with China, for every percentage of growth that China loses, South Africa could lose 0.2 percentage points.[6] And this is when one considers only imports and exports, and does not account for the effect of local conditions in South Africa.

There are two ways we can mobilise resources to invest in this economy: internal and external investments. It is said that South African corporations are sitting on a cash pile of R1 trillion, with cash reserves in the JSE's largest 50 companies increasing from R242 billion to R1.4 trillion between 2005 and 2016, as many firms do not invest domestically. According to the Business Leadership South Africa 2018 report, FDI had fallen from 24% of GDP in 2008 to 19% in 2018. The government, however, now aims to push FDI to 30%, in line with the National Development Plan (NDP). At the same time, the South African economy is very dependent on outside investors for its economic growth. For this reason, South Africa is highly concerned about how international investors view it.

Of course, a necessary step for South Africa is to ensure that there is an ease of doing business, particularly when it comes to business regulations and greater efficiency; this, in turn, will facilitate domestic producers as well as make doing business in our country easier for foreign companies.

The World Bank's annual Ease of Doing Business report, for example, which ranks countries according to the attractiveness and ease of doing business in a country, is determined by the complexity and cost of regulatory processes, as well as the strength of legal institutions in a country. While it is true that much of the FDI on the continent is still linked to the exports and imports of the coal, natural gas and oil industries, for instance, there needs to be a focus on improving the ease of doing business for there to be a further FDI injection. Yet, South Africa has significantly lagged behind here. In 2019, South Africa fell in the World Bank's Ease of Doing Business report to 84th out of 190 economies included in the study, slipping two places to its lowest ranking yet. While we held our rank in 2018 and 2017, we have seen a significant shift in the last decade. According to Trading Economics – a website focused on issues relating to macroeconomics and which collates information from official sources in various countries – macroeconomic models and expectations indicate that South Africa's ranking in the World Bank's report will improve to around the 69th position by 2021. So, while under President Ramaphosa, the government has made a concerted effort to improve the country's ranking in global business-competitiveness measures, much needs to be done in government departments to meet compliance and obvious backlogs. There has been some movement towards achieving this, but the existing efforts are simply not enough. In 2011, for instance, the Department of Trade and Industry launched the Companies and Intellectual Property Commission website, where individuals can register new businesses online.

When the Chinese paramount leader Deng Xiaoping was modernising China, he realised that for the country to succeed it would have to open up to the outside world. The story often told is that when the Chinese leadership was deciding on how to do this, Xiaoping took a pen and circled on a map the rural area of Shenzhen and declared it a special economic zone to serve as a magnet for outside economic investment. Also, Xiaoping realised that there was a need to capacitate China internally. He invested in education, which had been virtually closed for fifteen years during the Cultural Revolution. He also invested in strong leadership. In this regard, he stated that China's leaders should be younger and trained in technical subjects. Today, Shenzhen, with its twelve million people, is a US$640-billion economy – richer than South Africa, with a population of 55 million people. At the same time, South Africa is also introducing special economic zones, but the question is whether these will have enough economic gravity or innovation to attract investments and economically transform our people.

The first thing we need to do is to invest in our people. For us to be able to attract investments, whether domestic or foreign, we will need to overcome certain obstacles that stand in our way. The first is the crisis of high-level skills. To transform South Africa, we need technical skills. In this regard, according to the Engineering Council of South Africa (ECSA), one engineer in South Africa services 3 166 people, whereas in Brazil this ratio is one engineer to 227 people, and in Malaysia it is one engineer to 543 people. Furthermore, in South Africa, half of our trained engineers leave the profession for other fields, such as finance. Often, the argument is whether it makes sense to increase the number of engineers if they are not finding jobs.

This is a circular argument because without skills we will never produce sufficient jobs, whereas without jobs we will never have enough interest from our people to study these subjects. The lesson then is, whatever the circumstances may be, let us educate our people.

Mzala lived in a different era to ours – for one, there was no internet – but he understood objects in their totality and would have cautioned that, as we prepare for the 4IR, we should not forget to contextualise that within the broader framework of industrialisation. To paraphrase Mzala, he would have firmly understood that the 4IR should be cooked inside the broader industrialisation pot. He would have understood that the aim of any technological change is to modernise the productive forces. He would have understood that education is the only motivating force able to capacitate us to participate in the 4IR and eliminate poverty, unemployment and inequality. He would have understood Deng Xiaoping when he said: 'To uphold socialism we must eliminate poverty. Poverty is not socialism.' He would have understood that modernising our productive forces requires skilling our people to build scientific leadership.

CHAPTER 18

Leadership

On a 2018 visit alongside then Minister of Higher Education and Training Naledi Pandor to meet student leaders of universities and TVET colleges at UJ, I wondered whether amongst those student leaders there were pathfinders – those who dare to go where no one in their communities have been brave enough to go. Pathfinders are 'mafukuzela', because they are the ones who, despite obstacles, achieve their goals. I listened attentively to observe whether those students had the necessary logic and insight to tackle the problems of the current age as defined by the 4IR.

I wondered whether, amongst them, there were pathfinders such as Charlotte Makgomo Maxeke, who became the first black South African woman to receive a university degree, more than 120 years ago, in the US. I wondered if, amongst them, there were future leaders of the calibre of Frieda Bokwe, who became the first black South African woman to receive a university degree from a South African university, or John Tengo Jabavu, who founded the newspaper *Imvo Zabantsundu* in 1884.

As I listened to the problems of these youngsters, I wondered whether leadership in the 4IR should be any different from the leadership that has come before, in all of the other industrial revolutions. The first, second and the third industrial revolutions were about the centrality of people in the production of goods and services, and therefore required people to specialise. These revolutions required people to go to work and do as their bosses instructed. Africa was a spectator, not a participant of these industrial revolutions. Leadership in these revolutions was mainly limited to the ability to command and control the masses. Not so for the 4IR, in which the use of intelligent machines is the central ingredient, because it requires innovation, learning, analytics, agility, a global mindset, cultural tolerance, diversity, and the ability to manage conflict. In fact, according a 2018 *Harvard Business Review* article, in this era defined largely by AI, 'qualities, such as deep domain expertise, decisiveness, authority, and short-term task focus, are losing favor'.[1] However, other characteristics, such as humility, adaptability, vision and constant engagement, have emerged as the real keys to effective leadership.

The recipe for leadership

While the first three industrial revolutions were very destructive to the environment, the 4IR, according to Chinese President Xi Jinping, requires us 'to protect the environment while pursuing economic and social progress – to achieve harmony between man and nature, and harmony between man and society'. I wondered whether our student leaders understand industrial revolutions and their impact on people and the environment.

It became apparent to me that, due to the complexity of the problems humanity faces today, it is only logical that those who do not know should not lead. Historically, in South Africa, we have always been led by those with more knowledge and education. Leaders, such as Maxeke, Bokwe and Jabavu, paved the way for us to have the confidence to succeed. The legacy of Charlotte Maxeke lives on through the Wilberforce Community College in Evaton; the legacy of Bokwe lives on in the 2019 statistic that shows that, in South Africa today, more women than men are attending higher education; the legacy of Jabavu lives on in the numerous newspapers that serve African people today. But this is not enough, and it begs the question of whether our student leaders are ready to transform our society from a developing to a developed country.

In a report in 2019 titled *Leadership in the Fourth Industrial Revolution: Faces of Progress*, Deloitte found that as companies prepared to adapt to the 4IR, four types of leaders adapted more readily. The first set is termed 'social supers'. These leaders are defined by their ability to do good by implementing initiatives that have a positive social impact. The second set is 'data-driven decisives', who take a methodical and data-centred approach to decision-making, which includes leveraging many of the technologies of the 4IR. Here, using defined strategic processes and data to make decisions can create competitive advantages. The third set is 'disruption drivers', who focus on investment and disruptive innovation for a competitive edge. Much like data-driven decisives, disruption drivers make data-focused decisions based on input from multiple stakeholders. The fourth set is the 'talent champions', who prepare their employees for digital transformation by investing in the upskilling and retraining of employees. As Punit Renjen, global CEO of Deloitte, put it, 'As we look ahead, the qualities of leaders will define the future stability of organisations. Leaders who embody the characteristics of the successful leaders above are not only improving their bottom lines and growing faster than their counterparts but are also visionary in the ways they lead their companies into the future.'[2]

For us to succeed as a nation in the 4IR, we must be able to provide our people with education in its totality. This will mean that those who are interested in science and technology should also be required to study human and social sciences. Similarly, those who are interested in human and social sciences should also be required to study science and technology. Why is knowledge a crucial tool for the 4IR? Socially, knowledge liberates us from superstitious thinking and equips us to tackle complex problems. Are our student leaders receiving education in its totality to address complex issues?

When Deng Xiaoping was changing China after 500 years of lagging behind the West and the humiliation of being invaded by a relatively small country, Japan, he realised that education should be at the centre of his strategy. He advanced what he termed the 'four modernisations': advancement in agriculture, industry, national defence, as well as science and technology. In this regard, he reoriented the Chinese education system to be more international, modern and forward-looking. The results of these reforms have made China the most dynamic economy of our time. Are our student leaders ready to come up with theories and practices that will transform the economic landscape of our continent?

Why is this Chinese strategy relevant to South Africa, particularly as we enter the fourth industrial age? It is essential because, as we tackle the problems of unemployment, poverty and inequality, economic growth is crucial. It cannot be that an average Australian is ten times richer and more educated than an average South African. As we modernise and reform our agriculture through land reform, we must have sufficient people who are knowledgeable in the political economy of farming to ensure continued food security and global competitiveness. As we tackle the problems of our shrinking industries, we need to understand issues of automation, technology and human capital in order to improve the quality of life of our people. Any collective knowledge gap in our understanding of technology or of global economics or our knowledge of human-capital competency will derail our ambition of creating a developed society. Are our student leaders closing the knowledge gap that exists between us and the rest of the world?

Homing in on education

Going forward, what kind of education should we introduce in South Africa? To begin with, we need to consider starting education earlier than we do, by introducing compulsory pre-education. Masaru Ibuka, one of the two founders of the Japanese electronics company Sony, believed that the most crucial period of education is before the age of three. As we move forward, we also ought to foster creativity in our educational experiences. This way, we will have a population that is oriented more towards finding solutions than merely complaining about problems. We need to foster multidimensional thinking skills where our people can connect with machines and human beings equally. We need to promote the culture of discipline and, in this regard, revisit the issue of national service for the youth in all seriousness. Are our student leaders disciplined enough to tackle global competition?

Universities play a fundamental role in developing skills for future generations, with academia now navigating technological changes. In South Africa, we are tasked with the mammoth responsibility of growing our flailing economy, battered by years of mismanagement, and addressing the rising unemployment rate; South Africa standing as one of the most unequal societies in the world. In order to do this, we have to empower people, especially our youth, to be able to compete on the global stage, offering them skills development, people development, and research and development as fundamental drivers.

As the career landscape fundamentally changes, universities and researchers have a fundamental role to play. In 2017, alongside Professor Bo Xing, then associate professor at the Institute for Intelligent Systems at UJ, I wrote about the implications of the 4IR on higher education. As we put it at the time, 'Higher education in the fourth industrial revolution is a complex, dialectical and exciting opportunity which can potentially transform society for the better.'[3]

With the demands and challenges of the 4IR, there is now a move towards new, flexible, often-multi-disciplinary curricula that steer away from the traditional focus on predefined categories and types of learning. This requires strong and robust conversations on research: what are the questions that should keep us on edge; what are the focus areas for a university; and how do we reorganise ourselves?

Xing and I detailed the changes that higher education institutions need to undergo in order to adapt so as to keep up with the 4IR. What this initially takes is a stark and often tricky lesson on the current packaging of knowledge into modules and qualifications and the way that these are both taught and learned. This has to be revisited – in part, because higher-education institutions are preparing current students for a different work environment, but also because they are a platform for people to re-examine their skills. We are fundamentally reimagining what the learner looks like. Traditionally, learners have been young, fresh out of high school and enrolled in on-campus, lecture-based programmes. However, this is changing rapidly – as can be seen in the increase in the enrolment of older, working adults, who have unique learning requirements. The changing face of workplaces and the ways in which systems are shifting in our society mean that a particular task undertaken manually today can be rapidly replaced through the automation of processes tomorrow. This means that the community, as a whole, needs to embrace the concept of reskilling. To illustrate, the traditional roles of bookkeepers, for example, have become obsolete due to the automation of accounting processes. To keep up with the game, a bookkeeper would need to learn and master the different software and, in a sense, carve a particular niche for him- or herself. This is where education needs to focus our direction and energy. It is a major game-changer, especially since the focus in higher education has always primarily been on the provision of access to those who are exiting the school system and, at the other end, fostering postgraduate growth. But, with the 4IR, the rules have changed. There is now a very clear need for low-intensity training spanning diverse fields in order to keep up with the rapid changes in the world.

In the past, graduates were largely preparing for a specific profession based on predefined notions of what would be required to succeed in that profession. This, however, is not the reality we are faced with any longer. As workplaces and professions fundamentally change, we also have greater access to knowledge.

Part of the solution then is to shift the focus from teaching to learning, with emphasis on real-world problem-solving abilities and a multi-disciplinary approach to curricula that are more interactive. In tandem with traditional classroom learning, there is a move towards including student engagement through peer-to-peer interaction and one-on-one counselling that holds great promise for students. This form of cooperative education and co-construction of the teaching and learning processes is very much part of the 21st-century university. With the aid of technology, even a WhatsApp or Facebook group to discuss projects or assignments can be used to appeal to the learner of today. This is, of course, a reasonably basic usage of technology in the teaching and learning process. With the shift towards home-schooling and remote learning during the COVID-19 pandemic, many learners are connecting to the internet to continue their schooling. The 4IR can, however, be directly injected into the learning processes through AI, learning analytics and mobile-based learning platforms, which personalises the learning experience. For example, wearable technologies, such as those that make use of AR (VR headsets, for instance), have the potential to simulate real-life experiences. AR can supplement students' perception of reality via superimposing computer-generated information of their view of the physical environment in real-time, and this can help them to explore and interpret learning material in a more interactive way.

Many universities have already started working with industry by incorporating particular skills in their curriculum. There is also a substantial influx of employers who are now partnering with universities to improve tailored learning programmes for their employees in order to prepare them for emerging job opportunities. Part of this initiative includes relooking at how degrees are set up. While, up until a few years ago, there was only a little wiggle room in the subjects you could choose, this rigidity no longer works.

All these will be achieved only if we have dynamic leadership. We need to train leaders who can make rational decisions that are based on logic and facts. Rational leadership strives for efficiency and does not put resources to waste. Our leaders must be able to make effective decisions, even with incomplete and imperfect information. In my book, *Artificial Intelligence Techniques for Rational Decision Making*, I propose methods for making the right decisions, even in the presence of incomplete information. We need to train leaders who know how to listen and to interpret the message they hear, so that they can make decisions that benefit us all. These leaders must be able to negotiate a better deal for us, even in the presence of power asymmetry. Nobel laureates George Akerlof, Michael Spence and Joseph Stiglitz first proposed the idea of understanding and managing information asymmetry, and recent studies have shown that technology can reduce information asymmetry for better decision-making. We need to create an environment where efficiency, prudence and minimalism in consumption are inculcated in our culture. In this regard, the 'efficient-society hypothesis' – adapted from the 'efficient-market hypothesis' that we looked at in Chapter 16, wherein a system is considered efficient if it is rational and does not misallocate resources – must define our communities.

Our leaders must be connected to our people, technology and the world. They must be present in our communities, government and industries. They must learn and read because knowledge is extracted from reading and experience. It is, therefore, imperative that as educators, we offer these student leaders education in its totality and impart to them the skills to exercise leadership in the 4IR. In fact, Google even provides information in small chunks if required. A leader with a background in philosophy will have the skills, in the 4IR, to understand problems, create arguments, look at issues from multiple perspectives and posit solutions.

To build leadership in our people, it is essential to look for viable international examples. In England, prospective politicians study a degree called Politics, Philosophy and Economics. This enables graduates to be skilled in the art of political wheeling and dealing, to understand the economic and financial systems, and to have adequate philosophical training to understand the big issues of our times. The world has changed, and the leaders we train should understand politics, economics and technology. In this regard, UJ is introducing a Bachelor's degree in Politics, Economics and Technology to prepare our leaders for the 4IR. So far, UJ is the only university in the country to do so. This does not exclude other degrees from the mix, but rather highlights the need for thinking out of the box when planning higher-education degree programmes. The structural limitations foisted upon degree structures are largely archaic and work on assumptions of knowledge structures and boundaries that are slowly dissipating. Since the 1980s, we have been seeing a trend wherein engineers and medical professionals opt for MBA programmes. This melding of multiple fields and disciplines is precisely what is needed today. It can be argued that educational leaders should be looking hard at the school curriculum to understand what is obsolete and redundant, and what can be repackaged. To paraphrase Deng Xiaoping, our leaders must be innovative and must understand politics, economics and technology. The skills identified as a requisite for success in the 4IR must be infused in the pedagogy at universities.

PART 4

Society

It is not simply our jobs and current notions of work that are changing. There is a fundamental shift in our societies. Every industrial revolution has been accompanied by significant social change, ranging from shifts in language and political systems to how we entertain. This section delves into the nitty-gritty of the effects of the 4IR on society, beginning with creating an ethical framework and evolving to relying on AI systems to counteract human irrationality. It looks at the shifts we have seen in communication, in sports, in movies, in the way we work, in our democratic systems and in our ability to learn from history.

CHAPTER 19

Languages

In the throes of apartheid in 1960, South African singer Miriam Makeba turned the international spotlight on isiXhosa by introducing the 'Click Song' to the world. As she says at the beginning of a recording of the song, 'In my native village in Johannesburg, there is a song that we always sing when a young girl gets married. It is called the "Click Song" by the English because they cannot say "Qongqothwane".'

isiXhosa is an exciting language, with more than nine million speakers. For many, it can be a difficult one to master because, apparently, it has so many click sounds. Yet isiXhosa only has eighteen clicks compared to Jul'hoan, spoken in Botswana and Namibia, which has 48 clicks; or Taa, also spoken in Botswana, which has more than 100 click sounds. isiXhosa is, in fact, not a clicking language but a Bantu language. The American linguist Joseph Greenberg classified African languages into four stocks, one of which is the Bantu group spoken from Tanzania to South Africa. Bantu languages belong to the Niger–Congo group of languages. For those of us who speak a Bantu language, it is quite easy to pick up on the other Bantu languages. I speak Tshivenda, so isiXhosa is completely comprehensible. For example, 'malume' (meaning 'uncle') in Tshivenda is 'malume' in isiXhosa, while 'makazi' ('aunt') in isiXhosa is 'makhadzi' in Tshivenda. In addition, 'iza apha' ('come here') in isiXhosa is 'ida hafha' in Tshivenda. But, even though isiXhosa may not be that difficult for other Bantu-language speakers to grasp, it is very difficult for artificially intelligent machines.

Machine learning and Bantu languages

In Douglas Adams's *The Hitchhiker's Guide to the Galaxy*, when a small yellow fish, called the Babel fish, is placed in someone's ear, it allows them to hear any spoken language translated into their first language. While we are certainly not inserting fish into our ears, we have developed AI systems that can perform a similar task. Google, for instance, has developed quite a sophisticated tool known as Google Translate, which can translate words from one language to another. But, unlike with the Babel fish, not all languages are covered by this technology. It was only in 2017 that Google Translate extended its service to include isiXhosa, isiZulu and Swahili. Google Translate is based on deep learning, which allows it to translate spoken and written words into another language. Launched in 2006, Google Translate used United Nations and European parliament transcripts and documents to gather linguistic data. Rather than translating languages directly from one into another, Google Translate first translates the text into English and then into the desired language. Its machine-learning systems sift through millions of documents for patterns that help it decide what words to use and how to arrange them in a sentence. In a letter from the mathematician Warren Weaver to Norbert Wiener in 1947, Weaver famously said, 'When I look at an article in Russian, I say, "This is really written in English, but it has been coded in some strange symbols. I will now proceed to decode."' While deep learning can take words spoken in Chinese and translate them into English and vice versa, and while it can do this very well, isiXhosa – and most African languages – remains difficult for these machines to understand. Translating isiXhosa into English is often littered with laughable mistakes, with isiXhosa and isiZulu often mistaken for Indonesian. In 2017, a video of a parliamentary session with subtitles from Google Translate went viral. Not only did it mistranslate words, but it also could not pick up on South African accents.

For those of us who have been working in AI for over twenty years, this is an interesting challenge. Why is isiXhosa such a difficult language for machines? To answer this question, one needs to understand the classification of the Xhosa language. The people known broadly as Xhosa today – among them Nobel Prize-winners Nelson Mandela and Desmond Tutu – are a mixture of Nguni, which is a Bantu language, and Khoisan, the latter a people who speak a clicking language using the four-click Khoisan system. A study conducted by the HSRC found that, in terms of genealogy, of all African ethnic groups in South Africa, the Xhosas share the highest genetic likeness with the Khoisan. As it turns out, isiXhosa has the highest incidence of clicks among African languages in South Africa. This correlation between Xhosas and Khoisan, in effect, means that, of the linguistic and genetic cross-pollination that took place between the first nation of South Africa, the Khoisan, and other ethnic groups, that which took place with Xhosas was the most extensive. And yet, despite this close interaction between the Khoisan and the Xhosas, isiXhosa is not generally a clicking language.

When AI translates spoken words from one language to another, it takes a recording of the spoken words and breaks this down into individual sounds, tiny bits of audio that it can interpret, and, it converts this into a digital format as 'signals'. These signals are then decoded, or deconstructed, by the machine so that it can recognise them according to their 'cycles'. In 1822, the French mathematician Joseph Fourier was the first person to be able to come up with a method of understanding such signals, and he was thus able to comprehend that all signals can be represented as a combination of cycles – in mathematical language called 'sinusoidal functions'. The idea of a signal being represented as a combination of cycles was also observed by Karl Marx in his critical work, *Capital*. So, a word spoken in isiXhosa is deconstructed using the Fourier method so that it can be broken down into cycles that the AI machine is able to understand. When converting the signals into cycles, out-of-the-ordinary characteristics, such as those that the machine thinks come from background noise, for example, are eliminated. In the case of isiXhosa, the machine thinks that the click sounds are not part of the language, but rather background noise, and it thus eliminates them completely. This is particularly problematic when we consider that isiXhosa is comprised of 15% click sounds, and 85% Bantu language.

Another problem is that, often, AI-based translation software is predicated on the linguistic rules of Indo-European languages, which usually follow a predictable pattern. But in Bantu languages, the grammatical tense, references to a person, numbers and genders could be condensed into a single word rather than four distinct words in a sentence. So, much of the difficulty with translating isiXhosa comes from the fact that the language has historically been viewed through a Western lens. isiXhosa is too vital a language to be locked out of the 4IR.

Google Translate is, therefore, not perfect, especially as it does not obey Hillel the Elder's principle of reciprocity, which essentially states: 'Do unto others as you would have them do unto you.' For example, Google is more accurate in translating words from English to isiZulu than from isiZulu to English simply because the designers of this component of Google Translate know English better than they do isiZulu – which is, of course, contrary to the principle of reciprocity, where designers know isiZulu and English equally well.

Naturally, we need to fix this, ensuring that our entrepreneurs build and sell localised systems. Another limitation of Google Translate is its inability to translate complex expressions. For example, in isiZulu when one sends condolences, the phrase 'akwehlanga lungehlanga' is used, but Google Translate appears to suggest that this expression means 'race does not break'. Other examples are the isiZulu expression 'ukwanda ngomlomo njengembenge', which Google translates as 'oral increase as a banner' rather than as 'talking too much without any action', and 'ukushaya ngemfe iphindiwe', which Google translates as 'smoking is repeated again', instead of 'dealing with a person severely'. For Google Translate to be able to function well in our context, it requires isiZulu mother-tongue speakers to be involved. Better yet, our people need to create our own tool that translates from isiZulu to English and vice versa. And because my own mother tongue, Tshivenda, is not covered by Google Translate, I have often wondered whether I should build and sell a system that will translate any language into Tshivenda.

Google Translate is, however, a forerunner of devices that will be implanted in our ears that will take any words in any language and translate them into our native tongue. This will be the post-language era, when there will not be any need to learn any language for utility purposes.

So, the digital divide in Africa does not consist only of those who have access to technology and those who do not, but also encompasses the language divide. In 2018, American media company ComScore predicted that by 2020 half of all online searches would be voice-based, especially since voice-based systems, such as Amazon's Alexa, Apple's Siri and Google's Home assistant, have become increasingly common. While this has not completely materialised, we have indeed seen an increase in the use of voice-based searches. These devices are predicated on natural language processing (NLP); in other words, the ability of computers to comprehend the human voice and language. Through Alexa, for instance, you can shop online with a voice command, play a song, pull up your calendar or just chat. Yet, as this entire ecosystem of voice-prompted technology develops, there is still little done to make it applicable to an African context.

Google's investment in African languages is clearly limited. As the company has explained, essential criteria need to be in place first: it must be a written language and there need to be significant translations already available online before it can use a combination of machine learning, licensed content and the online translation community, in order to deliver its service.

It improves its service through a crowdsourcing model that asks users to provide feedback on the accuracy of translations or suggestions for alternative translations. However, this means that the less widely spoken a language is, the less its translations on Google Translate will be improved through such crowdsourced feedback.

Research conducted by the early-stage accelerator Digital Financial Services Lab and research consultancy Caribou Digital has also found that there is little incentive for these companies to include more languages. According to the research, published in *Quartz Africa*, 'by multiplying the number of speakers of a language by the gross domestic product per capita, the authors found that the top 100 languages cover approximately 96% of global GDP. Yet these 100 languages comprised less than 60% of all populations, highlighting a fundamental tension between the commercial and social value of languages'.[1] This, of course, is compounded by the fact that there is not enough data on African languages to train machine-learning systems – hardly surprising when you consider that African languages were often viewed through a reductionist lens while the focus remained on European languages. It was only as colonial legacies began to fall away that these languages were recognised and studied. For instance, Swahili – which is spoken by almost 100 million people across at least six countries – does not have an established repository that can be used to feed speech-recognition software, as English or French does. For these devices and services to pick up on languages, it is estimated that they require around 100 000 hours of recorded speech. So, while these obstacles exist as the primary challenges, there also isn't a single agreed-upon use of a single African language. Many dialects are different, and many use mixed-language models.

AI for African languages

What then is to be done? Firstly, we ought to develop a new translation technique that will not treat click sounds as noise but as an integral part of the language. One way to handle this is to fashion a new version of the Fourier method, one that can focus on the signals directly and not disregard the clicks. Another technique is to come up with new types of AI machines that will focus on the raw spoken words without any pre-processing and analyse them as an integral element of the language rather than discard them as noise. This will require us to discover new forms of algorithms that are decolonised and take into account the peculiarity of our languages, particularly difficult ones, such as isiXhosa.

What are some of the ideas that need to be explored? In the field of cognitive science, we have what is referred to as the 'cocktail-party problem', first described by Colin Cherry in 1953. Cherry observed that when you are in a noisy room, you can indeed hear the words of the individual you're talking to, which suggests that the human ear can actually filter out noise. So, in order to deal with the difficulty of the Xhosa language, one only needs to turn the cocktail problem on its head: not filter out the clicks conventionally deemed as noise by AI machines, but take them into account so that the AI machine can hear them, understand them and transmit them.

Fortunately, a great deal of work has already been done to ensure that AI machines understand the identity of the African people. One example is the work of Gugulethu Mabuza-Hocquet, who completed her doctorate in 2017 on designing algorithms that can understand the fact that the pupil-iris boundaries are not always perfect circles sharing the same centre, as is the case for people of European descent; in people of African descent, pupil-iris boundaries are different. These algorithms, therefore, allow biometric systems that use pupil-iris boundaries to identify people to not implicitly discriminate against Africans in favour of Europeans. The next step should be to develop better algorithms that understand the Xhosa language.

In 2019, Nigerian developer Gabriel Emmanuel began creating OBTranslate, a messaging platform capable of translating over 2 000 African languages. This was built using machine learning, AI and big-data analysis, which included identifying language patterns.[2] Some of its main uses include the translation of educational materials, television subtitles and song lyrics.[3] It not only looks at African languages but also picks up on pidgin English, which is a hybrid of English and local languages. The technology depends, however, on Africans who can teach these machines their local dialect.

Similarly, in 2018, Masakhane – meaning 'We Build Together' in isiZulu – was established as a cross-continent, open-source AI project to develop the neural machine translation systems needed to put African languages on the technological map and connect Africa's diverse and numerous linguistic populations.

The AI bias against Africa

As Tom Ilube, founder of the African Science Academy for Girls in Ghana, once put it, 'Algorithms define the future and people forget that algorithms are not just technical, they are political and cultural.'[4] The bias against Africa is not only apparent in Google Translate. Electronic maps have also become popular over the last few years. These maps can direct an individual from one location to another, and they are voice-activated. Their use has become so widespread that taxi drivers are said to be losing their spatial intelligence due to dependence on them. Google Maps is another popular electronic service that exploits Google search to allow people to move from one location to another without needing to know street names but, as we have already seen, these electronic maps are unable to pronounce our street names well. This is because the audio for these services is voiced in American and other accents as opposed to that of African-languages mother-tongue speakers. To change this, either the companies that make these maps must truly localise by employing first-language speakers or, as Africans, we should create our own service that can pronounce local street names.

Taking our languages into the digital and fourth industrial ages is our responsibility as South Africans and, indeed, as Africans. We cannot simply import technology, such as speech-recognition services; instead, we need to adapt them to our particular environments. If adaptation is not an option, we need to formulate our own versions of the Fourier theory. This will require our funding agencies, such as the NRF, to sponsor projects that are rich in local content rather than us having to solve the problems of other nations and thus subsidise them. This, in turn, would require a new sense of confidence and a realisation that the African market is big enough to define its own technological problems and solutions. Any other way will simply reinforce colonial economic, political, social and technological systems.

Now, how do we prepare Africa to exploit these business opportunities? Firstly, we need to be ready for the 4IR in terms of regulations, policies and investment strategy. This means that we should have enough people, as well as sufficiently developed physical and digital infrastructure, to help us make that transition. When it comes to infrastructure, the 4IR requires access to AI software such as TensorFlow. What we need to do is to get our people to study AI at primary, secondary, tertiary and adult-based education levels. Secondly, we need computational infrastructure that processes and stores data. We do not necessarily have to create this infrastructure – we can buy it. Companies such as Amazon and Microsoft are now selling local Cloud infrastructure with a capability for storage and Cloud computing. The CSIR's Centre for High-Performance Computing can play a role in the creation of such infrastructure. Thirdly, we ought to start collecting data in all areas of national importance. Data is the new oil in the 4IR, and so we need to invest in data-collection technologies. Africa also needs to create a law that makes data an IP asset. Fourthly, we should train people to have a deeper understanding of the inner workings of AI so that they are able to build relevant AI systems, collectively understand the problems in our society and use these to create solutions that advance Africa's business interests. Going forward, instead of waiting for multinational companies to include us in their business models, let us build our own businesses that will collaborate, through licensing and other mechanisms, with these companies.

CHAPTER 20

Ethics

In a documentary titled *Do You Trust This Computer?*, Tesla and SpaceX CEO Elon Musk offered a grim warning against superintelligence: 'At least when there's an evil dictator, that human is going to die. But for an AI, there would be no death – it would live forever. And then you would have an immortal dictator from which we can never escape.'

There have long been warnings about the technologies of the 4IR and the devastating impact they could potentially have. Much like in the movies of the 1980s and in many dystopian novels, the conundrum is whether the good outweighs the potential pitfalls. Dystopian novels have been littered with stark warnings about the downfall of humanity that will inevitably accompany technological change. As Aldous Huxley once put it, 'Technological progress has merely provided us with more efficient means for going backwards.' Huxley, who is most famous for his novel *Brave New World*, sketched an eerie picture of the future in the 1930s. Set in a futuristic World State, inhabited by genetically modified citizens who are categorised according to an intelligence-based social hierarchy, in *Brave New World* Huxley weaves a terrifying tale that prophesises scientific advancements in sleep-learning, psychological manipulation and classical conditioning that ultimately alter the way human beings think and act. At the end, the flawed hero eschews it all. Huxley feared that embracing scientific advancements would reduce human beings to passivity and egotism. And, in more recent years, it is not uncommon for TV series to explore these fears. In *Humans*, self-aware AI robots are at odds with humans over their consciousness. In an episode of *Black Mirror*, a woman discovers that technology now allows her to communicate with an AI that mimics her dead boyfriend. In *Westworld*, the robots are programmed to feel pain and fear but do not know they are robots.

Despite the caution espoused by many, however, AI is increasingly being used to perform tasks previously done by human beings. For example, doctors study electroencephalography (EEG) signal in order to detect epilepsy but, for a variety of medical, interpretive and social reasons, epilepsy has still often been misdiagnosed. An AI doctor, on the other hand, can perform the same process more consistently than a human doctor, who may be tired or biased. Despite the impressive results offered by AI, some serious ethical issues need to be tackled. Before his death, physicist Stephen Hawking warned that the emergence of AI could be the 'worst event in the history of our civilisation'. In January 2015, dozens of AI experts – including Stephen Hawking and Elon Musk – signed what some might consider a threatening open letter on AI, calling for more research on the societal impacts of AI. While they conceded that there would be significant benefits, they called for more investigation into any potential pitfalls. Alongside the four-paragraph letter, titled 'An Open Letter: Research Priorities for Robust and Beneficial Artificial Intelligence', they attached a twelve-page document of proposed research priorities. As the letter reads, 'The potential benefits are huge since everything that civilization has to offer is a product of human intelligence; we cannot predict what we might achieve when this intelligence is magnified by the tools AI may provide, but the eradication of disease and poverty are not unfathomable. Because of the great potential of AI, it is important to research how to reap its benefits while avoiding potential pitfalls.'[1]

What is ethical AI?

First, it has to be said, many of these fears are not entirely unfounded. While there are worst-case scenarios – that AI can be programmed to create AI weapons or develop destructive methods of performing otherwise-simple tasks – there are already misuses of digital and 4IR technology. For instance, in 2019, a video of Nancy Pelosi, the speaker of the US House of Representatives, was intentionally slowed by 25%, resulting in the pitch of her voice being altered, in order to make it appear as though she was slurring her words. The video, which went viral after it was initially posted on a Facebook page called Politics Watchdog, is one example of how technology can be used maliciously to manipulate and distort political messaging. Another is the use of deepfakes. A deepfake is an image or a video in which the face of someone, usually a celebrity, is pasted onto the body of another person so that it realistically appears as if the celebrity is doing what the original person in the image or video was doing. Deepfakes are created using AI.

This makes use of deep-learning technology – a branch of machine learning that is able to alter data sets – to create a fake. In this case, the technology learns what a face looks like at different angles and is thus able to transpose a face onto a target as if it were a mask.[2] While many amusing deepfake videos, as well as doctored videos like the Nancy Pelosi one, are available online, the use of them in politics presents a disturbing reality. How do we discern the fake from the genuine? The Nancy Pelosi video prompted US intelligence officials to issue a warning ahead of the 2020 elections about the use of deepfakes to influence political campaigns. This, of course, is just one concern. Could deepfakes be used for more nefarious purposes? Consider the implications of a deepfake video of a head of state announcing that a country would be launching a nuclear attack on another nation. In August 2017, Elon Musk stated that humanity faced a more significant risk from AI than from North Korea.

If you look at some of the biggest AI companies, they all have a set of rules and ethical considerations set out. Essentially, to avoid the robots taking over or engaging in a full-scale AI war, AI needs to be aligned with existing human values. As the signatories of the AI open letters asked: how do we create AI systems that are beneficial to society? Microsoft will tell you that its AI principles are predicated on fairness, inclusiveness, reliability, safety, transparency, privacy, security and accountability. Despite AI mimicking human thinking, humans still need to remain in control of AI so that artificially intelligent machines do what we want them to do. Yet, how do we ensure this and how do we agree on a set of ethical principles?

Perhaps the first step is to understand where the pitfalls lie. The consensus is that AI lends itself to bias, that there is a lack of accountability and transparency, and that there is not one set of human values for it to adopt. While there is an agreement that transparency, justice and fairness, nonmaleficence, responsibility, as well as privacy, are the core ethical principles, the meaning of these principles differs, particularly across countries and cultures. As tech philosopher Tom Chatfield explained in an article for Medium, there is a misconception that there is a set of AI codes for automating ethics that can be put into place once a broad set of codes has been established. As he puts it, 'Ethical codes are much less like computer code than their creators might wish. They are not so much sets of instructions as aspirations, couched in terms that beg more questions than they answer.'[3]

AI ethics, of course, goes beyond this. How do we put systems in place that also tackle the potential job losses and the genuine risk of exacerbating inequalities? As WEF executive chairperson Klaus Schwab warned, 'For all the opportunities that arise from the 4IR – and there are many – it does not come without risks. Perhaps one of the greatest is that the changes will exacerbate inequalities. And as we all know, an unequal world is a less stable one.'[4] According to research by McKinsey, AI will reinforce the digital divides that are already fuelling economic inequality and undermining competition.

So, what is to be done? One posited solution is to create a set of ethical guidelines within a set context. For instance, the Massachusetts Institute of Technology (MIT) has developed an open-source, middle-school AI ethics curriculum to make students in school aware of how AI systems mediate their everyday lives. This, of course, is applicable to a US context. The idea is not only to explain how AI is designed but also to illustrate to the students how it can manipulate them. It aims to prepare students for the jobs of the future. For example, students have to write an algorithm for the best peanut-butter-and-jelly sandwich. The debate centres around what 'best' means. Is it the best-tasting or the best-looking or the biggest? This demonstrates how bias is built into algorithms. In another activity, students are given images of both cats and dogs and use Google's Teachable Machines to create algorithms. Yet, the data given to the students is biased in that it includes more images of cats than dogs, which makes the AI system better at recognising cats than dogs. The students then discuss possible solutions, such as including more images of dogs, for instance, and alter the system based on those discussions.[5]

Ultimately, ethical AI has to be done in collaboration with all the stakeholders in society, ranging from government and the private sector to those in academia. According to Accenture, countries need to:

- set up an AI advisory body;
- collaborate internationally to develop comprehensive ethical AI codes;
- develop core ethical principles; and
- encourage the development of sector-specific codes of ethics.[6]

AI ethics and the law

Uber records revenue of over R80 billion per year. Companies such as Uber are known as managed services providers (MSPs). The value-add of MSPs lies in connecting customers with suppliers. Uber runs a taxi business, yet it does not own a single taxi. Because these MSPs operate on Cloud-based IT platforms, they can easily avoid local regulations and control. For example, when a customer catches an Uber in Johannesburg, he or she pays the taxi driver through a payment platform based in San Francisco.

Another consideration is that Uber has been making huge investments in self-driving cars – in other words, cars that do not need a driver because they drive themselves. This has implications for how our laws, designed for human drivers, will apply to such autonomous technologies. They drive on our roads and, therefore, are subject to our rules and regulations, such as speed limits. The question is thus: who is responsible for the fine if the self-driving car runs a red light or exceeds the speed limit? According to our laws, if a driver is caught driving over the speed limit, the driver, and not the owner of the car, is liable for the fine. So, what happens when a self-driving car goes over the speed limit? Given the fact that the vehicle drives autonomously of its human owner, do we charge the owner?

In Arizona, a self-driving Uber killed a pedestrian. According to preliminary investigation reports, this self-driving car observed the passenger but went ahead and killed her nonetheless. If this car had had a human driver, the driver would have been charged with involuntary manslaughter. But because this was a self-driving car, no one was arrested for the crime. For Uber to release this self-driving car onto the roads, the Chief Technical Officer (CTO) of Uber had to give permission. Is Uber's CTO thus liable for the alleged crime? Now is the time for the South African parliament to create laws that govern autonomous robots.

Suppose a self-driving car has four passengers and there is a pedestrian on the side of the road. If the car reaches a point at which it has to either avoid knocking the pedestrian by swerving and careening off a cliff, killing four passengers, or knocking the pedestrian and saving the four passengers, what should it do? Philosopher Jeremy Bentham came up with the theory of utilitarianism. If the self-driving car applies utilitarianism, it will do that which will bring the 'greatest amount of happiness to the greatest number of people'. So, if it saves the four passengers and kills one pedestrian, then the four passengers will be happy to be alive. If it kills the four passengers and saves the pedestrian, then the pedestrian (one person) will be happy. So, to keep the greatest number of people happy, the choice is clear: kill the pedestrian and save the four passengers. Now, if we are to move away from utilitarianism to ubuntu philosophy and the pedestrian is a three-year-old child, and the four passengers are all over the age of 60, then the self-driving car should take out the four passengers and save the child. It is clear from these scenarios then that our legislature needs to enact laws in South Africa that will ensure that these self-driving cars and any intelligent machines in our factories operate according to our values – values based on the principle of ubuntu. To do this, our legislators will need to understand the principles of AI and its implications. Our engineers will have to develop the capability to remodel these intelligent robots so that they are embedded with decision-making capabilities in line with our values.

Another ethical issue in AI relates to its use on social networking platforms, such as Facebook, Twitter and Instagram. The business models of these entities are based on the principle that they give you an account in exchange for your data. When an individual uses these programs, data on where she goes, what information she searches for, what clothes she likes and where she shops, for instance, is collected and sold to advertisers. When Mark Zuckerberg was asked about this in the US Congress, he replied that he believed that people should have the right to do what they want with their data. The problem with his response is that many people have no idea where their data ends up when they sign up for these services. In the case of South Africa, an added security issue is that much of the data collected locally is stored in California. The South African parliament should thus consider introducing a law that guarantees the basic right to privacy, which should be protected even in the event of a user agreeing to a social network's terms and conditions that might seek to alienate this right. This should be similar to how, under our law, the right to life cannot be waived through signing any kind of legal agreement.

One of the most significant inventions in biometric security is facial-recognition technology based on AI. Facial-recognition algorithms make use of databases that store countless images of faces, each of which is associated with the corresponding person's name. These systems are now on our phones. It turns out that the faces that are used to 'train' these AI machines are predominantly of Caucasian people, and those least represented in the databases are people of African descent. The consequence is that facial-recognition systems then discriminate against African people.

This, of course, is not ethical, and our laws will need to intervene to make sure that such discrimination does not persist. Our legislature will have to develop laws that will ensure that products – including this type of hardware and software – imported into South Africa comply with the country's Constitution and do not directly or indirectly discriminate.

Another significant ethical dilemma is the issue of clinical trials. A clinical trial is a process in which medicines and medical equipment that have been developed are tested on people to assess whether they work or not. A few years ago, Megan Russell, David Rubin, Brian Wigdorowitz and I registered a patent in the US for a 'robot voice'; one that would be used by someone who has had their voice box surgically removed because of cancer.[7] The result was that a big international medical-devices company took an interest in our patent and, during discussions, the issue of where these devices would be tested came up. Our international counterpart indicated that the laws surrounding clinical trials in their country are stricter than in ours, indicating that the trials would have to be done in South Africa, even though the device was to be sold in the international market. Studies have shown that Africa is becoming a home for clinical trials because the laws governing these are not strict – sometimes even non-existent. The African regulatory framework is, however, robust and should perhaps – through the Pan-African Parliament – help other African nations develop robust policies and regulate technology that will protect human lives and dignity.

CHAPTER 21

Democracy

When South Africa transitioned into democracy in 1994, it was a vastly different landscape to the one we have now. It was a watershed moment brought about by international pressure through sanctions, an unwavering movement within the country and the political will to stave off a civil war. In 1849, Henry David Thoreau asked, 'Is democracy, such as we know it, the last improvement possible in government? Is it not possible to take a step further towards recognising and organising the rights of man?' In the last few years, we have certainly seen a shift as the technologies of the 4IR and democracy interlink. While to some extent this has fostered more citizen participation, it has also been used as a tool in political campaigns. Writing on the 'Crisis of Anglo-American Democracy', Jeffrey Sachs asked how the world's two most venerable and influential democracies ended up with Donald Trump and Boris Johnson at the helm.

In his book, *Future Politics: Living Together in a World Transformed by Tech*, Jamie Susskind explains that technology has increasingly allowed us to share links and read news that confirms our views and beliefs while filtering out the information we do not agree with. Our newsfeeds on Facebook or Twitter are primarily tailored to our beliefs. For instance, Susskind explains that if you are a liberal who uses Twitter to follow the bi-annual election race for the US House of Representatives, 90% of the tweets you see will come from Democrats. 'Problematically,' says Susskind, 'this means that the world I see every day may be profoundly different from the one you see.'[1] Coupled with the prevalence of fake news, the technologies of the 4IR have created increasingly polarised groups. This begs the question: how is the 4IR shifting democracy, and is it a threat to it?

South Africa's democratic transition

When I was born in Venda in 1971, three years after the great partition of the peoples of Venda, I was registered as a South African citizen. The great partition was the separation of Tshivenda speakers from their Tsonga and Sepedi counterparts, who had been living together for many generations. The Tsonga speakers were moved from the hinterland of Venda to what is now called Malamulele. To create Malamulele, the Kruger National Park fence had to be moved. Sepedi speakers were moved from the Sinthumule area to Botlokwa, Bandelierskop and the other regions outside present-day Polokwane deemed reserved for the Bapedi people. This was to comply with the notorious Group Areas Act, which prescribed that only people who spoke the same indigenous language should live together.

This separation, which saw the movement of 10% of people living in Venda, was nothing less than a disaster. Firstly, you could hardly tell who was Venda, Pedi or Tsonga because these people had intermarried so much that the division was, at best, superficial. These three groups even established a language called Tshiguvhu, which was mutually comprehensible to all three language groups. Then the gender question entered the great partition! A person's ethnic group was defined by the ethnicity of their last name, and family names were of course determined by patriarchy. Then began the separation of families! Brothers and sisters were assigned different ethnic identities depending on whom they married and were thus separated on a great scale.

The Bapedi were lucky because the places they were taken to were not as affected by malaria. Consequently, many of them survived, even though they were greatly impoverished because they still had to start again without compensation. The Tsonga speakers were, however, less 'fortunate' because Malamulele was a malaria area, and 10% of the people perished following the move – a forgotten genocide following the Group Areas Act.

In 1979, eight years after my birth, I lost my South African 'identity' and, at the stroke of a pen by those who processed my reapplication, I became a citizen of the Republic of Venda. I still have my identity document from the Republic of Venda, which was established after the republics of Transkei and Bophuthatswana. The irony of the story is that my last Venda ID was issued after 1994 – after Nelson Mandela became president of South Africa. This was not malicious in intent, but simply because the 'homelands' had not yet been integrated into the new South Africa. The other irony is that even though I was a citizen of Venda, I held a South African passport because the

Venda passport was only acceptable to a handful of countries, among them Taiwan, which some believe is a province of the People's Republic of China. Our transition to democracy was thus a historical event, but it was painful and full of ironies.

Democracy in the 4IR

How do we guard against the negation of our democracy through the technologies of the 4IR? Because we spend so much time on social media networks, such as Facebook and Twitter, and communicate using so-called free email systems, such as Gmail, large portions of our society are being watched, studied and nudged to behave in a certain way. This is industrialising the way our politicians interact with us, and so they rely on technology rather than face-to-face interaction. Twitter and Facebook, for instance, use AI-powered algorithms to tailor your feed based on your interests, desires and preferences. While the algorithms were initially created to streamline the massive amount of information available, the Facebook algorithm ranks available posts by how likely you are to have a positive reaction to them. As internet activist Eli Pariser puts it, 'Left to their own devices, personalisation filters serve up a kind of invisible auto-propaganda, indoctrinating us with our own ideas, amplifying our desire for things that are familiar and leaving us oblivious to the dangers lurking in the dark territory of the unknown.'[2] The same, of course, is valid for Google. As Cathy O'Neil remarks in *Weapons of Math Destruction: How Big Data Increases Inequality and Threatens Democracy*, Google's search algorithms are focused on raising revenue, but it also has a dramatic effect on what people learn and how they vote. This principle was tested by researchers Robert Epstein and Ronald E Robertson, who asked undecided voters in the US and India to use a search engine that was skewed in order to learn about the elections. They found that this shifted voting preferences by 20%.

The proliferation of fake news also poses a threat to democracy. According to an MIT study, the top 1% of fake news reaches between 1 000 and 100 000 people, whereas the truth rarely reaches more than 1 000 people. It is becoming increasingly simple to spread fake news. In 2019, fake news further fuelled xenophobic attacks in South Africa. As President Cyril Ramaphosa stated, at least ten people were killed, of whom only two were foreigners; whereas news of higher death tolls and gruesome accounts of violence flooded social media. This has had a detrimental impact on democracy.

For example, in India, fake news increased by 40% ahead of the parliamentary elections, while in Nigeria, fake news about violence in polling stations went viral. This is becoming increasingly difficult to track. While it can be reported on Facebook and Twitter, fake news can spread unchecked on platforms such as WhatsApp or Telegram. We even saw during the #FeesMustFall and #RhodesMustFall campaigns that it was difficult to distinguish between truth and falsehood. In fact, in some cases, even the media was using social media posts as sources for their articles or radio and television programmes, without checking the veracity of information. This has seen the emergence of organisations, such as Africa Check, which focus on establishing the truth of information put forward.

In 2019, prior to the national elections, I received a pre-recorded call from a leader of one of our major political parties urging me to vote for their party. It turns out that these political parties use AI software that accesses social media data of users in order to segment undecided potential voters, and it contacts them directly to nudge them to vote in a particular manner. The idea of harvesting data and using it to achieve a political objective is exactly the scandal that engulfed Facebook and Cambridge Analytica. It was not only in the election of Trump that Cambridge Analytica played a role. In 2019, it emerged that the data-analytics firm also worked for the 'Leave.EU' campaign, as well as for the United Kingdom Independence Party over the 2016 referendum on EU membership. It is clear then that we have to guard against 4IR technologies taking away our political independence and ultimately our vote. As former US Secretary of State Henry Kissinger put it, 'If the convictions expressed by the candidate are the reflections of a big-data research effort into individuals' likely preferences and prejudices, then what once had been substantive debates about the content of governance will reduce candidates to being spokesmen for a marketing effort.'[3] This form of aggressive marketing happens unchecked and, despite some reporting mechanisms, the legal system within the country is yet to contend with the aggressive strategies used by political parties. It must be said that modern-day political parties employ similar sales and marketing tactics customarily used by corporates. Freedom of speech in a democracy has thus found full expression through social media platforms, including the usage of bots created for specific causes.

The 1994 democratic transition in South Africa was more evolutionary than revolutionary, but it gave us a democratic vote. Democracy is precious, but its value depends on how much we invest in making it work. If we do not make it work, it will be characterised by what is known as the 'banality of democracy', a term borrowed from the concept of the 'banality of evil', coined by German-American philosopher Hannah Arendt to describe how easy it is for ordinary people to descend into evil. When it comes to the notion of the banality of democracy, it is astonishing how easy it is for people to take democracy for granted, simply by not becoming informed voters. In the crowded spaces of social media, information and misinformation can go viral very rapidly. This often makes it difficult for citizens to discern truth from falsehood.

As a historical materialist, I believe that the measure of the effectiveness of democracy is how much it is able to solve the contradictions of unemployment, inequality and poverty that plague our society. Democracy on its own will not solve these problems. There has to be a strong feedback mechanism between the voters and elected officials so that officials are warier of disappointing the electorate. For us to strengthen this feedback mechanism, there has to be a strong connection between what our elected officials do and what is good for the general population, and, for that to happen, we need informed and not banal voting.

How do we achieve informed voting? We need to educate ourselves about the most pressing issues affecting our communities and communicate these to our public representatives, whether local, provincial or national. For example, I argue that only a tiny fraction of people in our population know where their parliamentary constituency offices (PCO) are located. Worse yet, very few even know what the PCO is, and the names of the elected representatives assigned to their local PCO. Whenever there is a serious problem in the community, instead of first going to their PCO, which is what generally happens in a functional democracy, the tendency is to protest. Protesting is important, but protesting without holding our public representatives accountable is meaningless.

This is not to say that AI cannot also be beneficial for democracies. Based on an understanding of individual preferences, AI can increase voter participation. The automation of elections can curb fraud and corruption and leaves little room for human error in the counting of votes or the rigging of elections. The company Camatica has called on the South African government to use its AI-powered facial-recognition technology for the municipal elections in 2021 in order to eliminate issues around the lack of ballot papers, to avoid any inappropriate behaviour of election officials, and to improve the flawed system of marking thumbs with a permanent marker. AI could also be used to keep politicians accountable by triggering a new set of elections whenever confidence drops – although this, of course, could prove challenging in contexts such as South Africa where business and consumer confidence have reached historical lows despite the elections. AI also has the potential to create a fairer justice system that would eliminate the need for human judges and deliver more impartial court rulings. For instance, in 2008, Judge John Hlophe was referred to the Judicial Service Commission for 'an improper attempt to influence this [Constitutional] Court's pending judgment in one or more cases'. This, of course, would not be a worry with an AI system. In order to hand down verdicts and sentences, judges make risk assessments. This is based on various factors, such as the evidence, how this was presented by the attorneys and how witnesses testify. AI systems, on the other hand, generally make more rational and optimised decisions. Here, judgments can be made by taking into account the evidence and, for example, the defendant's history and come up with a recidivism-risk score – which estimates whether they are likely to reoffend – in order to determine a sentence, for instance.

AI superpowers

But what happens when democratic freedoms are threatened? As China looks set to emerge as the first AI superpower, a debate around the AI state and democracy has emerged. While the Chinese government claims it runs a socialist market economy and the Chinese Communist Party (CCP) is the government, this political system does not necessarily encompass many of the aspects of democracy as we understand it. For example, criticising the Chinese government on social media will result in you losing points in your social score on the Social Credit System. China's use of data, AI and internet surveillance seems to threaten our notions of democracy. Harvard historian Julian Gewirtz once said, 'When the Chinese government saw that information technology was becoming a part of daily life, it realised it would have a powerful new tool for both gathering information and controlling culture, for making Chinese people more "modern" and more "governable" – which have been perennial obsessions of the leadership.'[4] Yet, China is leading the charge when it comes to the 4IR, so how do we reconcile that with the gaps in its democracy? As Steven Feldstein puts it, 'Around the world, AI systems are showing their potential for abetting repressive regimes and upending the relationship between citizen and state, thereby accelerating a global resurgence of authoritarianism. China is driving the proliferation of AI technology to authoritarian and illiberal regimes, an approach that has become a key component of Chinese geopolitical strategy.'[5] Are other countries able to compete with China when it comes to collecting data on this scale or using it for digital surveillance?

Already we see other countries replicating parts of China's AI strategy. Singapore has, for instance, put state-of-the-art facial-recognition cameras in place as part of its surveillance programme, with 110 000 cameras fixed to lampposts around the city state, while Zimbabwe is producing a national image database for use in facial recognition.[6] As Feldstein says, 'As governments become increasingly dependent upon Chinese technology to manage their populations and maintain power, they will face greater pressure to align with China's agenda.'[7]

The question then becomes: how do we ensure that democracy is not threatened? The examples above demonstrate the authoritarian usage of AI, but we must consider that governments can use AI especially in countries where access to information, education and health care is restricted, based on economics or geographical distance from the centre. It would require robust investment in infrastructure to provide government and the people with the necessary capabilities to obviate challenges in the provision of access. For example, in South Africa, access to complex specialised care for those in rural areas can be enhanced by AI. This could assist with diagnosis and treatment, and simultaneously reduce reliance on our overcrowded tertiary hospitals. In the field of education, schools with limited access to fully-fledged laboratories could access, through multimedia platforms, material about how experiments are conducted; they could also perform dissections using AR/VR, and learn about technology through technology. The truth is that governments are dealing with diverse populations with dissonances in terms of inequality and inequity, shrinking fiscal resources, and stark tensions between equipped urban cities and towns versus sparse infrastructure in rural areas. The solution can be found in utilising AI, thus having a positive impact on society. In South Africa, where the tensions of service delivery are rampant, surely the government could track progress and make services available in smarter ways.

It may be surprising to naysayers who view the advent of the 4IR with scepticism that social-justice agendas can be pursued with rigour, advancing causes at the forefront of socio-political agendas. What is clear is that AI cannot be driven merely by the tech crowd; it also needs activists, those from other disciplines, to ensure that the usage or introduction of technology does not encroach on human rights. This is because the algorithms used in AI are not free of bias. Recently, we saw a medical-aid company using big data to avoid payouts to specific segments of society based on race. In a democracy, the use of watchdogs to guard against this is thus critical. Appropriate governance of AI and technology can avert any subversiveness that may be contradictory to the ideals of democracy. The tangible usages of AI can, therefore, complement the workings of democracy by enabling in-depth analysis of data to inform policy development, track policy implementation and measure consequences and impact. If at the heart of a democracy is the notion of serving the citizens, then AI – with the required safeguards – surely provides the capabilities.

CHAPTER 22

Movies

Movies have long been associated with AI. We can scan back as far as 1927 to the first instance of the portrayal of AI in a film, the first of many such portrayals that would give rise to a rather erroneous understanding of AI. The 1927 silent German film *Metropolis* is about a humanoid robot intent on conquering the city. AI continues to get a bad rap in films. As it turns out, however, movies have gotten much of AI wrong and have contributed to much of the fearmongering. Yet, at the same time, our science-fiction writers and science-fiction films have also portrayed technological developments often decades before they have come into fruition. Oscar Wilde argued that life imitates art more than art imitates life, and before digital cinematography, moviemakers were able to imitate life, and the future, by using physical special effects, which were captured on celluloid film stock. The character R2-D2 from the original *Star Wars* films was one such example of the portrayal of an intelligent machine long before the automated age.

The latest instalment of the *Terminator* film franchise is another example of how far we have advanced, technologically. The latest movie seems a little more frightening than the *Terminator* movies shot in the 1980s. Back then, robots taking over was the stuff of science fiction and far removed from reality. The original film's plot involves a cyborg assassin being sent back in time from 2029 to 1984 to kill Sarah Connor, whose son will one day become a liberator against machines in a post-apocalyptic future. By then, in 1984, we had already seen automation in the previous three revolutions: mechanical automation in the first industrial revolution; electrical automation in the second industrial revolution; and electronic automation in the third industrial revolution. These advances meant that machines took over labour-intensive jobs from humans, but that people discovered other, more cognitive jobs to do instead. Yet, when Arnold Schwarzenegger returned to screens in 2015 – just as he promised he would – it painted a dark picture of AI. Enhanced human cyborgs run riot in the new *Terminator* – a fear for many when you bring up AI, and it is even scarier than before because now it seems like a distinct possibility.

Proponents of AI have long laughed off the portrayal of technology in films. You can find lists on sites, such as Science Mag or Wired, dissecting AI in movies. Where did they go wrong? What did they get right? But perhaps the real conversation we should be having is how these technologies are actually changing the way movies are made. Following the Oscars at the beginning of 2020, a debate raged on about whether AI could create an Oscar-worthy film. After all, AI was recently used to create an artwork considered such a masterpiece that it sold for US$432 500 at British auction house Christie's. An Oscar for an AI-generated movie does not seem quite so far-fetched.

AI-created movies

Alfred Hitchcock's famous thriller *The Birds*, made in 1963, focuses on a chain of swift, mysterious and violent bird attacks on the people of Bodega Bay, California, over the course of a few days. This was not an easy feat pre-computer-generated imagery (CGI), a technology that makes use of computer graphics to create 3D images and special effects in both live-action and animated movies. Hitchcock used real, live birds rather than mechanical ones. Special effects were then overlaid on this initial footage. The effect of the flapping of the birds' wings was done in the Walt Disney Studios by an animator using the sodium-vapour process known as 'yellowscreen' that combined actors and background footage. While this was momentous, technology has evolved significantly since then.

In 1993, when the first instalment of the *Jurassic Park* franchise was released, it was the first time CGI graphics, in the form of the dinosaurs in the movie, shared the screen with human actors. This heralded a revolution in film production that allowed for the creation of entire CGI cities or landscapes. CGI is now one of the most common tools deployed in the film industry. Some of the most lovable movie characters for children are sophisticated animations with human-like qualities, which have found universal appeal. *Toy Story*, for example, was the first feature-length computer-animated film.

And, more than two decades later, there has been a further breakthrough. In 2019, it was announced that CGI would be used to bring James Dean back to life in digital form for a Vietnam War movie called *Finding Jack*. Using thousands of images and footage of Dean, machine learning will create a virtual version of the Hollywood star that can be used in movies and video games. The announcement of this project was not, however, well received in Hollywood. The initial backlash came from actors. Chris Evans, who has played the character of Captain America in the Marvel-franchise movies of the same name, tweeted, 'I'm sure he'd be thrilled', with an emoji rolling its eyes. 'This is awful. Maybe we can get a computer to paint us a new Picasso. Or write a couple of new John Lennon tunes. The complete lack of understanding here is shameful,' he went on to tweet. Other actors shared the sentiment. There are, of course, further ethical considerations, such as using a person's likeness after their death. Robin Williams, for example, set a restriction on the use of his image, or any likeness thereof, for 25 years after his death.

While it is new to recreate an actor for a whole movie, the technology has been markedly used over the last few years in Hollywood. After the actress's sudden death during filming, CGI was used to complete Carrie Fisher's scenes in the 2019 movie, *Star Wars: The Rise of Skywalker*. Robert de Niro was de-aged in *The Irishman* using similar technology. In *The Lord of the Rings*, the character Gollum was created with CGI based on the facial expressions, body movements and voice of actor Andy Serkis. In *The Parent Trap*, CGI was superimposed on a body double of Lindsay Lohan so that it would appear as if she were Lohan's twin. This, of course, is just one aspect of filmmaking.

In 2016, a short film titled *Sunspring* was written entirely by an AI bot named Benjamin. Benjamin, the first automatic screenwriter, is based on a neural network known as long short-term memory (LSTM) often used for text recognition.[1] Benjamin was trained to write the screenplay based on data it gathered from dozens of others it found online, mostly sci-fi movies from the 1980s and 1990s.[2] Benjamin dissected them down to the letter and learned to predict not only what letters would follow each other, but also what phrases would occur in both stage directions and dialogue.[3] The premise of the story is that three individuals – H, H2 and C – are entangled in a web of murder and romance set in a futuristic world. While the experiment was ultimately successful, there were some mishaps. For instance, Benjamin was not able to create names because the sequence of letters in names is not as predictable as it is in other words. Instead, the characters were assigned letters. Some of the stage directions were, however, incomprehensible: 'H standing in the stars and sitting on the floor', for instance – as was the dialogue in some instances. In one scene, for example, H says, 'In a future with mass unemployment, young people are forced to sell blood.' C responds with: 'You should see the boy and shut up. I was the one who was going to be a hundred years old.' The man then proceeds to vomit up an eyeball. What this nevertheless demonstrates is the capability of machine learning in a creative context and, despite a few hitches, that AI could successfully execute the task of writing a screenplay.

Benjamin, however, has not been the only experiment. In 2019, a one-minute advert for Lexus was written entirely by AI that had sifted through fifteen years' worth of luxury ads. More recently, a documentary titled *They Shall Not Grow Old* was made using First World War film footage. Using adjustments to timing, the addition of colour and 3D imagery, the documentary begins in black and white and slowly transitions into colour.

Using AI to predict the next big thing

It is not only within movies that AI has been used. In the last few years, Hollywood has even been using AI to decide what kind of movies to make. AI can predict whether a film that is unsuccessful in the US may have a better run in Europe. It can predict whether changing the lead actor or actress would see it gross more money. For instance, Los Angeles-based Cinelytic analyses historical data about movie performances and then cross-references it with information about films' themes and critical talent. Producers can then input a cast or swap an actor for another to see how the movie might fare at the box office.[4] For instance, you could see if replacing Matt Damon with Leonardo DiCaprio might make for a more successful film or whether perhaps adding Leonardo DiCaprio alongside Matt Damon could be the key. Cinelytic lets you map out all the possibilities and how these permutations might change the film's box-office performance. Churning movies out for box-office success rather than for art itself, however, has implications in terms of quality and art.

This kind of analytics is also used to inform product development across multiple industries. Companies that are planning product launches have even successfully used it, with variations, for marketing purposes.

The company has compared the process of how it identifies the best actors to play certain roles to fantasy football, the game played online in which people can assemble a virtual team of famous, current football players and earn points based on their players' real-world performances and stats. Similarly, Cinelytic assigns scores to actors based on their past box-office performances or their social media profiles. This, of course, is not infallible. As Steve Rose wrote for *The Guardian*, 'The data these algorithms are processing are actually human beings, which are inherently erratic, fallible and unpredictable. Your asset might go into rehab, get divorced or decide to go off and make shoes for a year.'[5] Rose uses the example of Robert Downey Jr, who was arrested in 1996 for erratic behaviour, including driving his Porsche naked. As Rose put it, which algorithm would have cast him as Iron Man? – a move that has proven to have paid off based on the success of the *Iron Man* and *Avengers* franchises.[6] There are also concerns as to whether filmmakers will still take creative risks if the algorithm suggests otherwise, predicting failure at the box office, for example. We have already seen a move towards this as filmmakers opt for sequels or remakes rather than creating something original.

Similarly, ScriptBook, which was developed in Belgium, can predict a movie's success by analysing the script. Israeli start-up Vault AI can predict which demographics will watch their films based on how trailers are received online. Even established players like 20th Century Fox are using AI to project how a movie might perform. There are, of course, some flaws. While ScriptBook greenlit the horror *Get Out*, it vastly underestimated how successful it would be at the box office, predicting US$56 million in revenue. The movie went on to make US$176 million. What is apparent, though, is that Hollywood and its counterparts are using AI to predict success and identify variables that could, if modified, impact on the success of the film. Box-office takings are central to the business model of the film industry, so it is par for the course that analytics would be an integral component in decision-making.

As ScriptBook founder Nadira Azermai told *The Guardian*, 'All of a sudden you have this conservative, traditional industry versus companies who are believers in data. There is a war on content, but one party is using the latest technology, and the other is riding a donkey.'[7]

This type of predictive technology can also be used to nudge viewers in specific directions or predict future views. This is used by streaming services such as Netflix and Hulu, which have arisen alongside the shift towards using big data. These companies use algorithms to give you suggestions on what to watch next based on your viewing patterns – what kind of movies and series do you usually gravitate to? This recognition of a viewer's patterns is similar to the advertisement suggestions that pop up on Facebook or Google based on searches, comments, or posts that you make. If you recently watched *Black Panther* and *Avengers: Endgame*, it may suggest *Captain Marvel* or any of the *Batman* movies. This is not the only data they use. Netflix estimates that only 20% of its subscriber video choices stem from a search, with the other 80% coming from recommendations.[8] It knows which thumbnail teaser will convince you to click on a movie and even the choices you might make in interactive, choose-your-own-adventure style movies, such as *Black Mirror: Bandersnatch*. In these kinds of interactive titles, you can make choices for characters that shape the way the story plays out. As Netflix's head of product innovation, Todd Yellin, told the *Financial Review*, 'We have one big global algorithm, which is super-helpful because it leverages all the tastes of all consumers around the world.'[9] In fact, Netflix estimates that its recommendation algorithm is worth US$1 billion. Netflix's global algorithm makes recommendations by finding similarities among users in 190 countries.

Technology in set production and costume design

Technology is also fundamentally changing how films are made. 3D printing, for instance, has increasingly been used by prop makers. For example, in the Netflix show *Stranger Things*, 3D printing was used to transform the Demogorgon, the monster from an alternative universe, from its digital model into a physical form. As the *Stranger Things* art director and visual-effects supervisor Steffen Reichstadt put it, 'It's more accurate, and it saves money … 3D printing allows for a more precise build and gives a real-world perspective of form, weight, stature, etc. Things that are often overlooked in a purely CG workflow.'[10]

In *Black Panther*, 3D printing was used to create custom-made costumes based on mixing traditional African culture with new-age technology. Using 3D printing allowed costume designers to push the limits of design.[11] In *Star Wars*, the stormtroopers' helmets and some parts of the robot, C-3PO, were made with 3D printing.[12] In 2012, Laika produced *ParaNorman*, a stop-motion animated movie using only 3D-printed models. Before 3D printing, filmmakers had to hand-sculpt and paint every facial expression made by a character, but the makers of *ParaNorman* were able to 3D-print around 40 000 faces with different expressions.[13] More recently, Disney has been using robot acrobats to perform impossibly difficult scenes previously done by body doubles. Now, robot acrobats can dive off skyscrapers or be run over by trains, for example.

Similarly, filming movies has also evolved. As Randy Scott Slavin, founder of the New York City Drone Film Festival, told Wired, 'It's like when you look at your iPhone, it has more production capabilities in your little handset than NBC had in the 1990s – they surely couldn't shoot in HD and edit and all that other stuff without millions of dollars' worth of gear.'[14] For instance, smartphones, GoPros and drones are all being used to film movies. Selena Gomez's music video for 'Lose You to Love Me' was shot entirely on Apple's iPhone 11 Pro, which can take 4K video at 60 frames per second. Drones are also being used for aerial footage, which has since become increasingly common in movies. This has proved cheaper and more efficient than using helicopters for similar footage. For instance, in *The Wolf of Wall Street*, drones were used to shoot a party scene from above. But drone technology in films has also evolved beyond just capturing aerial shots. For example, in *Chappie*, the camera on the drone was used as the point-of-view shots of one of the robot characters.[15] In the James Bond instalment *Skyfall*, high-speed drone footage captured the opening motorbike chase scene.[16] Drones have also transformed the viewing experience in 3D and 4D films (the latter are 3D movies that are complemented by physical effects, such as vibration, motion and temperature changes, which are synchronised to the movie). To imitate the movement of the flying reptiles in *Jurassic World*, a drone-mounted camera pounces low over a crowd of people being attacked by pterosaurs. Watching this scene in 3D feels more realistic to viewers.

There is much that is invisible to the eye in the mysterious world of films. It seems clear that there are multiple components to the very complicated process of making movies, and the advent of AI means that the new capabilities are being used in innovative ways. One such instance is in the field of market assessment; another is using AI technology to steer, nudge, pre-suggest and prompt viewers on platforms such as Netflix. On the creative side, from the wonders of *Batman* to *Jumanji* to *Jurassic Park*, to name a few, there is a concerted use of highly sophisticated digital tools in the creation of sets and illusions. There is some evidence that points to the technical side of film forging forward in a bid to garner audiences and create the next wow moment, thus ensuring that this area will continue to see innovation breakthroughs. As *The Guardian*'s Steve Rose sums up this entire shift in the movie industry: 'Sounds like a plot for a great, metatextual Hollywood conspiracy thriller. If we got Dwayne Johnson, it could be a smash. Let's crunch some numbers.'[17]

CHAPTER 23

Work

If you are like me, you may strike up a conversation with your Uber driver, particularly on a long commute. Often my conversations with Uber drivers delve into how they became a driver and how business has been. Two decades ago, Uber was an unfathomable concept, but now you would be hard-pressed to find someone who has not used the app. From some Uber drivers, I have learned about how the world of work is fundamentally changing. The benefit, many tell me, is that working for Uber allows you flexible hours. This has been particularly helpful for many who work additional jobs or for those who need to tailor their day around their kids, for instance. In fact, this concept has created an entirely new economy – the gig economy. In my conversations, I have often been surprised to learn that my Uber drivers constantly double up variously as electricians, bartenders, photographers, freelance graphic designers and fashion designers, with one driver even having designed an entire men's fashion line.

This has become the new normal for many and is but one instance of how the world of work is changing. It is not uncommon, for example, to find out that your colleague is also an Airbnb host. Platforms such as Uber and Airbnb, among others, have cropped up with the digitisation of the economy, which has created a space for platforms based on technology that respond to demand. In South Africa, SweepSouth connects users with domestic workers for either an on-demand clean or a recurring booking.

This, of course, has not been the only shift in the world of work. The 4IR has fundamentally changed the very nature of many jobs too. As COVID-19 hit the globe in 2020, many governments declared national states of disaster, while the World Health Organization termed it a pandemic. In response, many countries went into partial or full lockdown. Suddenly, with the promotion of self-isolation, social distancing and bans on business travel, many were forced to work from home. Some countries saw total shutdowns, yet in a fragile global economy, where work could go on, it simply had to. This was perhaps the starkest example of the future of work.

As Arthur Goldstuck, tech analyst and founder of World Wide Worx, said, 'The coronavirus crisis will have at least one positive outcome. It will provide a dramatic, global and unavoidable case study of the fourth industrial revolution in action. We will quickly discover that 4IR is not about artificial intelligence and robots taking our jobs, but about digitally enabling much of the human workforce.'[1]

Yet, while this has quickly become the new normal for many, entire industries have also come to a standstill. How do we interrogate this in a context such as Africa, where many are unemployed, and many of those who are employed are unskilled and do not have the kind of jobs or capacity to work remotely?

Tackling the unemployment beast

Unemployment is a scourge in Africa. As we talk of the gig economy and the future of work, we have to look at how this plays out on the continent. While Africa has the youngest population, with 200 million people aged between 15 and 24, it also boasts a disheartening unemployment rate. Based on data from the African Development Bank (AfDB), the continent is faced with the challenge of having to create twelve million additional jobs, with youth unemployment sitting uncomfortably at 60%.

We also need to interrogate the kind of work available to those looking for jobs. A study by Brookings in Washington found that '[young] people find work, but not in places that pay good wages, develop skills or provide a measure of job security'.[2] As the institution puts it, this often masks the reality of countries with lower unemployment rates. In 2016, the International Labour Organization (ILO) found that up to 70% of African workers were actually 'working poor'.

In South Africa, the issue is particularly stark. In the third quarter of 2019, the unemployment rate climbed 0.1 of a percentage point to 29.1% – the highest level on record. This does not even encompass our discouraged workers, who have given up looking for work, which would put the estimate closer to the 40% mark. This ranks as among the worst in the world. When compared to other BRICS nations, South Africa's unemployment rate is particularly dismal. It is more than double Brazil's 11.6% and worse than India's

8.5%, Russia's 4.6% and China's 3.6%. Youth unemployment is particularly frightening in South Africa, with the figure sitting at 58.2%.

The future of work in South Africa

If we are to consider that a 2009 Accenture study concluded that approximately six million jobs in South Africa are at risk of disappearing because of automation by 2025, we have to relook at the world of work. It is not just limited to blue-collar jobs; white-collar jobs are under threat too. A study by Brookings renewed these fears, finding that the rise of AI is five times as likely to displace college graduates than those without a degree.

According to a report published by the National Economic Development and Labour Council (NEDLAC) – an organisation that is made up of business, labour and government representatives – 'The speed and depth of technology adoption drives business growth, new job creation and the need for augmentation of existing jobs, provided it can fully leverage the talents of a motivated and agile workforce that is equipped with skills to take advantage of new opportunities.'[3] The report imagines what the world of work will look like in South Africa by 2030, based not only on the 4IR but also on the South African context of deep-seated inequality, a dwindling economy and employment policies, among other factors.

While there will likely be a decrease in roles such as telemarketers, legal clerks, rental clerks, cashiers and tellers, there will be an increase in roles such as AI and machine-learning specialists, human–machine integration coaches and experience-focused travel agents. The report outlines, too, how individual sectors will be impacted. For instance, in health care, there is a shift towards overall well-being in the long term rather than just curing disease in the short term. In the energy sector, there is added pressure to control climate change by stemming carbon-dioxide emissions. The transport sector is already adopting technologies, such as drones, hyperloop – a sealed system of tubes through which a pod can travel without air resistance or friction at high speed – and autonomous vehicles. As outlined earlier, the public sector is also changing, with a renewed focus on service delivery using e-services. Mining will see the adoption of drone applications, proximity sensors and improved communication systems that are changing underground mining, which will not only make mining safer but also more efficient. Importantly, there will be an overhaul across every industry as technology infuses into the workplace.

A study by McKinsey has estimated that around one-fifth of the global workforce will be impacted by the adoption of 4IR technology, particularly in developed nations, such as the US, the UK and Germany. It was estimated that by 2020, half of all companies would have decreased their full-time staff; and by 2030, some 800 million workers will be replaced with robots. This, however, is not quite the job bloodbath you would imagine. In fact, it represents a shift in the world of work. As former tech analyst Vinnie Mirchandani explains, 'Machines will become more of our colleagues, and we should not be so worried about their increased presence in the future … [If] anything, they will take our outstanding workers and make them even better.'[4] Research from the WEF indicates that 38% of businesses believe that this shift will merely change the role of employees as robots supplement their roles, while 25% think this will result in the emergence of entirely new functions. This has already been evident. A decade or so ago, becoming a social media manager, or an influencer, would not be seen as a viable career path. Now, we are seeing brands move away from traditional forms of advertising towards using influencers. Research by IZEA Worldwide, an online marketplace that connects brands and publishers with influential content creators, shows that 56% of respondents in a survey made a purchase based on a sponsored social media post from an influencer. In comparison, 63% said that they found influencer-created content more compelling than traditional advertising.[5] Brands are also becoming increasingly aware of the need to be active on social media. Responses to significant events, interaction with consumers and brand recognition have become inextricably linked with Facebook, Twitter and Instagram. Nando's, for instance, has become somewhat infamous for pushing boundaries with controversial content on topical issues.

The notion of our working lives is also fundamentally changing. In the past, individuals would often work for a company for their entire career, with long-service awards testament to the decades spent at an organisation. Yet, this is not how millennials are tackling the workplace. Now, the reality is that many will change jobs and even careers multiple times. Although a decade ago, this may have been frowned upon and would emerge as a red flag for companies and recruiters, this mentality is also changing. This has actually made new hires more adaptable in the 4IR. For instance, individuals who have changed jobs and careers are often more likely to embrace the need for upskilling. Yet, there are companies that have not cottoned on to this idea. Research from Indeed, an employment-related search engine for job listings, found that 65% of employers surveyed said that they had opted not to interview applicants with short-tenure jobs at other companies. In comparison, 44% of employers felt that applicants having three short-tenure jobs on their CV amounted to job-hopping. If we are to promote the re-education of our employees, and expect them to adapt to the 4IR and have experience across fields, we must be more open to the notion of job-hopping as a positive rather than a negative.

In the 4IR, working remotely could become the new normal. Virtual private networks (VPNs) allow you to more securely log onto systems outside of a company's premises; apps, such as Facetime or Skype, allow you to have remote as opposed to face-to-face meetings, and WhatsApp allows you to stay constantly in touch with colleagues. Banks, such as FNB, have also encouraged a concept called Uberisation – in other words, the ability to work from anywhere – thus subverting the traditional understanding of how banks work. The popular notion of a banker is usually of a person in a stark black suit, wearing a hat and working in a very rigid environment. This is no longer the case. Today, there is room for more flexible hours and the option of working from home in casual clothes. While there are differing definitions of Uberisation, in FNB's case, this is not quite the gig economy that applies to freelancers or contractors but instead promotes the idea that work does not need to be confined to a particular office space.

It is not just the use of mobile phones and laptops that have made this a possibility; there are also other technologies that can be deployed to make this more effective. For instance, Microsoft Teams was developed as a shared communication and collaboration platform for video meetings, document collaboration, file sharing and storage. Citrix Workspace is software that allows multiple users to remotely access and operate Windows desktops via PCs, tablets and other devices. Similarly, TeamViewer can be used for desktop sharing, file transfer and online conferencing. Slack is an instant-messaging platform, designed specifically for organisations, on which messages can be shared across channels, and private messages can be used to share sensitive information or documents. Trello can be used to track progress, tasks and collaboration, which not only helps in terms of delegating work but can also aid performance management, which is more challenging to keep track of remotely. Many of these companies have offered their tools for free during the period in which the coronavirus has been running rampant, in the hope that it will become a standard once the COVID-19 pandemic has eased. As Matt Mullenweg, CEO of WordPress and Automattic, which owns Tumblr, told *The Guardian*, 'These changes might also offer an opportunity for many companies to finally build a culture that allows long-overdue work flexibility. Millions of people will get the chance to experience days without long commutes, or the harsh inflexibility of not being able to stay close to home when a family member is sick.'[6] In fact, in 2017, Nicholas Bloom, a Stanford Economics professor, conducted a study using a Chinese travel agency by the name of Ctrip to find out whether working remotely would increase efficiency. For six months, half of the employees worked from home, while the other half came into the office. The study found that those who worked from home were 13.5% more efficient and 9% more engaged than those who remained in the office.

The gig economy

The gig economy works around temporary positions or contracted work, the type done by freelancers, contractors and temporary hires, for instance. This is, in part, due to the fact that the workforce is becoming increasingly mobile and work can be done from almost anywhere. This does begin to provide some solutions to the unemployment crisis in the country and the continent. However, much of the workforce won't benefit from technological advances because they are unskilled or manual workers, which are among some of the first jobs that will likely be automated in the 4IR. In fact, Africa already boasts over 300 active digital platforms, which together employ nearly five million people. Yet, as research from the Fairwork Project found, 'Growing numbers of South Africans find work in the gig economy, and digital platforms are frequently heralded as a solution to mass unemployment … The employment challenge facing South Africa is not simply the number of jobs but also the quality of jobs being created.'[7] Of course, this is not the only worry in South Africa. The gig economy calls for some level of tech-savviness and financial resources. How does one tap into this economy when you do not have a smartphone or access to the internet?

The gig economy, of course, is not confined to ride-sharing companies, food couriers, or even the likes of Airbnb. A report by McKinsey & Company found that the fastest-growing segments of the freelance economy were from knowledge-intensive industries and creative occupations. And there are benefits for companies too. Not only do they save on office space and training, but can also contract experts they would not ordinarily be able to hire. This gives employers a larger pool of candidates from which to choose. For instance, a freelance journalist in Ghana could readily write for a publication in the UK. Similarly, an advertising agency in South Africa could rope in a copywriter from the US. Interestingly enough, some full-time employees are part of the gig economy and freelance in their spare time to earn extra outcome. The McKinsey study found that around 64 million people in the US and EU use gig work to supplement their primary income by choice rather than necessity.[8]

Of course, this kind of work structure has not been without its challenges, and the COVID-19 pandemic has already demonstrated weaknesses in the gig economy. In countries where there was a complete lockdown, such as Italy, Uber drivers have not been able to make any money whatsoever. This is because workers in a gig economy are not paid a regular salary – they are paid only for each gig they do. While food-delivery services, such as Uber Eats or OrderIn, have continued operating, this has not been without risks. In a gig economy, where companies that hire gig workers do not offer them benefits, many of these workers do not have health insurance or a safety net. Companies have been forced to rethink these policies. For instance, Uber indicated that it would compensate workers required to self-isolate for fourteen days. It is important to remember, however, that it is not only amid a pandemic that these gaps exist. There is also sometimes the worry of where the next gig is going to come from. In economic downturns – and, as demonstrated by COVID-19, during pandemics – uncertainty is a thinly cloaked veil over the gig economy.

As the world of work fundamentally changes, the COVID-19 pandemic has provided an interesting study of whether these models will work. Will working remotely be here to stay? How do we begin to address the gaps in the gig economy? Do we create legislation to protect those who work gigs rather than full-time jobs?

CHAPTER 24

Rationality

I come from Limpopo Province in the land of Tshivhasa, the paramount chief of a region north of the Tropic of Capricorn and south of the border between South Africa and Zimbabwe. In our area, there was a middle-aged man, John, who had a wife by the name of Sarah. The two had a son, James, who must have been about twenty years old. John was estranged from Sarah, and one morning they had an argument. Later in the day, it rained, and there was a thunderstorm. Lighting struck their house, killing Sarah. A few days later, in broad daylight, James came back and viciously murdered his father. We were there to witness this gruesome murder. Of course, the police arrived, and James was arrested and given a long prison term. Here are the facts: John and Sarah argued. Lightning struck Sarah. James made a causal link between the argument between John and Sarah and the lightning striking Sarah. In the act of murdering his father, James stated that John was responsible for creating the lightning. Of course, Sarah, John and James are fictitious names but represent real people I knew. When this happened, I was the tender age of ten and did not know the principles of electromagnetism that would help me understand the source of lightning. I know now, of course, that John was innocent. Making false causal links based on some hidden and often supernatural force is known as superstition. Superstition is irrational and can be very dangerous. As the philosopher Hypatia said, 'Men will fight for a superstition quite as quickly as for a living truth – often more so, since a superstition is so intangible you cannot get at it to refute it, but truth is a point of view, and so is changeable.'

Years later, I enrolled at the University of Cambridge to study for a doctorate in AI. The house where I lived overlooked the Parker's Piece common, the two separated by a road known as Park Parade. Every year there was a bonfire at the park. (Coming from the land of the Tshivhasa, I know something about fire. When the Tshivhasa came to our area, they burned people's houses to become the ruling chiefs of our region. The name Tshivhasa means 'the one who burns'.) The annual bonfire at Parker's Piece has been happening for over 300 years. Initially, it was to burn witches, and many innocent souls were burned at the stake at Cambridge. A significant number of those accused of witchcraft fled to the Americas to escape being burned. Unfortunately, however, the witchcraft label followed them. Burning people for any reason is immoral; burning people because they are thought to be witches is irrational and superstitious. As Christopher Hitchens, author of *Letters to a Young Contrarian*, said, 'Beware the irrational, however seductive. Shun the "transcendent" and all who invite you to subordinate or annihilate yourself.'[1]

Moving away from superstition

Of course, superstition is entrenched in communities across the world, not just in Africa and, in many ways, are intrinsic to our varied belief systems. However, we do need to guard against superstition driving behaviour or beliefs in ways that are harmful. For example, burning someone because you believe they might be a witch is not predicated on scientific knowledge. However, if you were to burn two red chillies to ward off evil spirits, this is not harmful – although, of course, for some, it would perhaps be irrational. Here, education is key, and we must call upon people to interrogate their superstitions. At the core of superstition is irrationality, which is directly opposed to rational thinking, which is the use of information and logic to arrive at a conclusion efficiently.

When I was appointed as vice-chancellor and principal of UJ, I outlined one of my missions to create a society that makes decisions based on evidence using scientific principles. My mission is to move our people from superstitious to scientific thinking. This is vital if we are to tackle the intractable problems of poverty, unemployment and inequality. China is an ancient society with a sophisticated civilisation, and, yet, until the 20th century, women's feet were bound so that they could not grow beyond three to four inches – a feature deemed to be more attractive. The result, however, was that women suffered severe deformities of their feet, had regular infections and struggled to walk as a result of disability. This cruelty was irrational. When the People's Republic of China was established, the practice was banned. Once Deng Xiaoping began modernising China, one of the principles he adopted was that people with scientific education should rule China. This explains the fact that the current president, Xi Jinping, is a chemical engineer, having succeeded Hu Jintao, a civil engineer, who in turn followed Jiang Zemin, an electrical engineer. Scientific thinking is so embedded in the thinking of Chinese leaders that Hu Jintao's theory – written into the Constitution of China – is known as 'scientific development'. The idea behind this theory was to move China away from an over-reliance on cheap and unskilled labour and industries to skilled workers and industries focused on science and technology. What Deng Xiaoping was doing was moving the Chinese people from superstitious thinking to scientific thinking.

In the era of the 4IR, people and machines are becoming a single being. Today, people cannot be separated from their technological gadgets. One of the modern instruments of torture is to separate a person from their smartphone. After a few hours, they may begin to experience withdrawal symptoms, which, according to studies, appear to mimic the brain activity of individuals withdrawing from narcotics.

Today, computer gadgets are so powerful that they can measure our temperature and heart rate, for instance. They gather so much information from us that businesses are being built around this use of data. The Discovery Health app, for example, which collects users' data, is essential in the actuarial pricing of health insurance. The technologies of phones today thus help us make rational decisions. As the proponents of Nudge Theory, Daniel Kahneman and Richard Thaler observed, human beings on their own are relatively irrational when making decisions. The pair further illustrated how such irrational thinking creates inefficient markets that are bad for fair trade. It is clear then that this irrationality is pretty much universal, stretching from the Tshivhasa region of Limpopo to Cambridge and Beijing.

Rationality

Given the pervasiveness of human irrationality, then, how can we use AI to temper this and help us make more rational decisions? One way is to create AI machines, or intelligent models, that use data to arrive at rational conclusions in an efficient manner. However, we need to decide on the best way to do this and to ensure that these machines are as efficient as possible.

When we create AI machines to make decisions for us, we use data to train them. The data has a dimension. For example, if we are building a machine that uses the temperatures at 12:00 in the last two days to predict what the temperature will be at 12:00 on the following day, then this data set uses two variables to predict one variable. In this case, the dimension of the input data is 2, and the dimension of the output data is 1.

When one builds intelligent models such as this, the complexity of the model is defined by how many free variables are used in it. The more variables used, the more complex the model is, and the most preferred model is the one that is the least complex. This is called 'Occam's razor' and is named after William of Occam.

When we model a system, not all input variables are equally relevant. Some will be marginally relevant. Including more variables than needed often leads to a phenomenon known as the 'curse of dimensionality'. The inclusion of marginally relevant variables makes the model unnecessarily complicated, which compromises the effectiveness of the model. The trade-off between the dimension of the model and the complexity of the model gives an efficiency frontier. When constructing intelligent models, then, one needs to decide how efficient it needs to be and where it will fall on the efficiency frontier. Therefore, one needs to align the dimension of the model with the desired complexity of the model. This choice between model dimension and complexity is subjective. If this model is used for rational decision-making, it is bounded, because the model is not perfect, and the rationality of this model is subjective.

We need to be aware, then, that while the systems we design may be rational, they are subjective. Machine rationality is subjective for three reasons, the first being that optimisation, whether for single or multiple goals, is subjective. The second reason is that the trade-off between model dimension and complexity makes rationality subjective. This trade-off between dimension and complexity does not just occur when intelligent machines make decisions but also when human beings make decisions. When humans make decisions, the more variables they use to make decisions, the less effectively they can process such variables. The fewer the variables they use to make decisions, the better they can handle the information. The choice that humans make on how many variables they use to process information versus the brain-processing effort, which is a subjective choice, ultimately makes bounded rational decisions subjective. Thirdly, training machines using data originating from subjective humans produces subjective machines.

Creating rational AI

With the advent of AI, can rationality be maximised? In research on whether AI machines are rational, it has been observed that, just like human beings, machines have limited or bounded rationality. What was also observed is that AI machines are more rational than human beings. Because of this, AI machines are now deployed in hospitals to assist doctors in diagnosing diseases, are currently working alongside pilots, used in our banks to assess loan applications, price assets and manage risks, and in drones to minimise military casualties. But it is evident that when deploying AI machines, we ought to decide whether humans are in or out of the loop in decision-making.

If we take the examples used in previous chapters, there are clear instances that show that machines are more rational. For example, in the game Go, Google DeepMind's AlphaGo shows that algorithms make better decisions. According to Ray Dalio, author of *Principles: Life and Work*, criteria for these kinds of decisions must be noted. As Dalio puts it, once an algorithm has proven to work under certain circumstances, then we should always use it. We, of course, have to also take into account AI bias. While humans may have similar biases, they often do not act on them. In 2018, Amazon had to scrap an AI recruiting tool that showed bias against women. The tool reviewed résumés with the aim of mechanising the search for top talent, but, by 2015, the company realised that bias had been built into the system, based on the dominance of men in the tech industry. The system taught itself that male candidates were preferable, and it thus penalised résumés that included, for example, the word 'women's', as in 'women's chess club captain'.[2] Similarly, AI has shown racial bias. Despite this, for decisions that are not as politically and socially charged and are already based on large quantities of data, AI is decidedly more rational. Given the fact that AI machines are, in most cases, more rational than humans, what is to be done to increase aggregate rationality in society? Firstly, we need to invest in AI technologies in our schools, communities and industries. Secondly, we should study the science of human–machine interactions in all facets of our society. Thirdly, we must embed universal scientific thinking in our preschool, primary, secondary and tertiary education. Fourthly, we should mobilise our society – whether in the political, civil or industrial spaces – to adopt technology as an essential motive force to solve our problems. Fifthly, and finally, we need to make a concerted effort in all our spaces to move our people to scientific thinking.

CHAPTER 25

Sports

In 2003, Michael Lewis wrote a book titled *Moneyball*, which was later made into a movie starring Brad Pitt in the lead role. Written nearly two decades ago, Lewis's book tells the story of how the general manager of the Oakland Athletics used data to create a competitive baseball team. While this would seem somewhat rudimentary now as we progress further into the 4IR, it was an essential precursor of data-driven performance optimisation. The story tells how Billy Beane, the team's general manager, adapted baseball analyst Bill James's statistical concepts of analysing the individual statistics of players to assemble a team that would go on to have its best season yet, rising unexpectedly up the rankings. James had found that using the on-base percentage, which measures how frequently a batter manages to at least reach first base every time he or she bats, was much more effective in evaluating a batter's effectiveness than simply using the batting average, which only measures the frequency of times the batter manages to hit the ball.[1]

In contrast, using the slugging average, which measures the average number of bases a batter reaches every time he or she bats, was more effective than the conventional method of only considering stats relating to counting the number of stolen bases, as well as bunts (which are when the batter gently taps the ball instead of swinging at it, in order to make it difficult to field).[2] The team, despite having a smaller budget than most other teams, went on to have a twenty-game winning streak. This was more than a decade before algorithms – as we understand them now – had been used to analyse hordes of data to achieve a similar outcome. As Lewis put it, 'People in both fields operate with beliefs and biases. To the extent you can eliminate both and replace them with data, you gain a clear advantage.'[3]

In fact, data has permeated the field to such an extent that, in 2019, AI was used to create a new sport. As senior writer Charlotte Edmond reported for the WEF, Speedgate merges concepts from croquet, rugby and soccer. In this game, teams of six players kick a rugby ball around a field with three vertical pairs of six-foot posts, known as gates.[4] Players must pass the ball through a gate in the middle of the field in order to gain possession, and they score by passing the ball through gates on either end of the field. Teammates who catch the scoring ball and immediately kick it back through the posts convert the goal from two points to three.[5] The game was conceptualised using data from 400 popular sports, which was fed into a neural network and then crunched to create a basic framework of rules and concepts.[6]

It is clear then that the 4IR is fundamentally changing sports for both players and spectators. In understanding the skills required in the 4IR, we have not only gained a new understanding of the importance of sports, but also seen a change in the way sports are played and analysed.

In our current digital age, with the sophisticated use of technology, the very nature of how sports are played, coached and managed is driven by data. The use of equipment that enables data transmission allows for granular levels of detail to be recorded and analysed, enabling both players and coaches to zone in on how to eliminate injuries, what strategies to use in specific conditions and, of course, to study detailed post-mortem analyses on factors that lead to a win or a loss in a game. While in the past this was done retrospectively, this data can now be a live feed that enables quick analysis to inform possible steps or actions to be taken. Noting the importance of physical performance and health of a sports player, organisations invest in technology for ongoing monitoring of an individual as well as a team in real-time. In fact, the technology and data generated are so deep and extensive that they can also be used by those with decision-making powers to determine the composition of a team, based on the metrics available on players. In competitive sports, where the stakes are high, and there is a global competition, the construction of teams based on historical data that includes physical performance, for instance, is already widely used. Tests used to determine emotional intelligence and strength now form part of a battery of diagnostic testing used to determine the composition of a team and ways of harnessing talents in individual ways. Advanced technologies enable simulation of a game that will take into account elements such as climatic conditions and a player's skill set. Most technologies can be used across the globe, with modifications for particular metrics or language. Since the days of the rudimentary stopwatch, sports and technologies have formed an alliance, and this is particularly so in the 4IR. Professional sportspeople endorse new high-end equipment, such as the Fitbit (wearable tech that monitors a user's physical-activity levels), and this usually helps to bring about the common adoption of these technologies.

Developing 4IR skills through playing sports

When I was a teenager, my father was a diehard supporter of Moroka Swallows Football Club. I was accustomed to him spoiling me with those bright-maroon shirts emblazoned with the message, 'Don't follow me, follow the birds.' Why anyone should follow the birds instead of their hearts remains a mystery to me, and I so hated those shirts that I stayed away from sports completely. Later, when I was an undergraduate student studying Mechanical Engineering in the US, I was compelled to take two sports classes. I took the easy way out and chose body conditioning and bowling.

According to the WEF, the top ten skills that will be required in the 4IR include people management, the ability to coordinate with others, emotional intelligence, good judgement, decision-making and cognitive flexibility. Many of these are difficult to embed in any curriculum at university, but sports can play a big part in developing these skills. For example, team sports require players to coordinate with others and to manage people. The concept of winning and losing, a fundamental element of competitive sport, develops emotional intelligence. A change in team strategy midway through a game improves cognitive flexibility, judgement and decision-making.

In medical research, there is a well-established relationship between sports and wellness, with sports having been proven to reduce stress and improve mental well-being. Team sports improve social skills and build a culture of cooperation and leadership, thus improving individuals' emotional intelligence, which has been proven to be a better predictor of success than intellectual intelligence. Studies in the US demonstrate that 95% of the CEOs of Fortune 500 companies played competitive sports in college: Facebook CEO Mark Zuckerberg was captain of the school fencing team, and Microsoft CEO Satya Nadella played cricket, for instance.

Through its wellness programme, Discovery Health incentivises its members to be physically active, simply because there is a link between physical activity and wellness. Studies have shown that the more physically active a person is, the lower her or his medical-aid claims will be, so incentivising members of a medical-insurance plan to be active is a good economic decision. The importance of physical activity even extends to the workforce. The more physically active the employees, the more productively they perform. Discovery utilises a downloadable app to track the health and performance of an individual, for example, by monitoring the person's spending on healthy foods. There is a trend these days for companies to link user data from wearable smart devices, such as a Fitbit, for example, to the rewards they will receive for healthy behaviour. In the case of Discovery Health, these rewards are scored using Vitality Miles. By allowing users of the app to share their data with their friends and connections, and by ranking performance on a scoreboard, the company motivates individuals to be competitive.

Given all the benefits of sports in terms of wellness and social skills, how do we nudge discouraged athletes like me who were intimidated by maroon shirts to start exercising? Firstly, we need to embed sports in our university curricula, making these compulsory for all students. The courses should not be credit-bearing, but should nevertheless be a requirement for a student to proceed to the next year. Secondly, universities should hold health walks in which both staff and students participate, and perhaps prizes should be given to those who complete these walks. Thirdly, universities need to invest more in sports facilities, which are one of the most visible signs of excellent universities simply because there is a recognised link between physical vitality and educational outcomes. The lesson we can learn from the rewards programmes of companies such as Discovery is that it is important to offer an incentive; even something as simple as rewarding users with a badge on their profiles in a smartphone app for completing 10 000 steps can be effective. It is a truism that skills learned in sport are transferable to the real world. As in Speedgate, it is teamwork that enables a win. To win in a game of rugby, for example, all players on the team have to fulfil their roles, and in golf, the discipline required to master the golf club is one that is valued in our current age.

Bringing the 4IR to sports

Arguably, the sports industry was one of the first to be entirely infiltrated by technology, ranging from wearable health-and-training devices to goal-line technology in football (which is an automated system that checks whether the ball has crossed the goal-line or not). In some cases, this has helped athletes up their game, while in others it has improved the impartiality of referees who can now catch rule infringements more accurately.

During the FIFA Women's World Cup in France in 2019, we witnessed the extraordinary role technology can play in the use of video-assistant referee (VAR). From the banks of the River Seine to the foot of the Alps, games were delayed for minutes as VAR footage was replayed in real-time, sparking a fierce debate around whether the last goal was offside or not, whether a subtle brush against a hand was a handball or not, and whether a red card was warranted or not. This has continued in the English Premier League games, with even fiercer controversy; the scrutiny and close attention to detail making the stakes higher than ever. The result is the elimination of bias or oversight. For example, if analysed using today's technology, Maradona's famous 'Hand of God' goal of 1986 would easily be ruled out through the use of sophisticated software. This doesn't necessarily mean that we should replace all human referees with robots, but it does mean that there is a valuable role for such technology to play in ensuring decision-making is fair and accurate. At Wimbledon and most other tennis tournaments, final judgement calls on whether a ball is in or out is made by AI-enabled technology. The jury is out as to whether robots entering the field of sports and taking the human element out of it would be a crying shame for all, the question being: should we be sacrificing the human element for accuracy? We have already seen players at Wimbledon berate the little sensors that beep when they pick up what the naked eye may not see.

Another example of the application of AI in sports is the use of wearable performance-monitoring systems that measure heart rate, breathing rate, heart-rate variability, posture and the impact of physical contact. Such a device can alert the user as to whether they have exercised beyond a safety zone and thus reduces injuries, optimises performance, facilitates return-to-play, as well as monitors player movements during training and matches. The collected data is then analysed using an AI system that recommends the best exercises for a given profile. At a more finessed level, coaches can remotely monitor the daily patterns of the performance and health data of their players.

Extending the application of 4IR technologies in sport to everyday life

It isn't just athletes, coaches or sportspeople in general who utilise various forms of AI in order to enhance their physical performance. For example, ordinary people also use devices, such as the Fitbit, which use AI to make health recommendations based on their heart rate, sleep patterns and activity levels.

Another area in which 4IR sports technologies have become useful in daily life is that of driver-tracking analysis. For example, Discovery uses technology that measures the driving performance of their car-insurance clients in order to reward them for good driving.

This kind of technology is most effectively and intensively used in the sport of Formula One (F1) racing. F1 vehicles are fitted with between 200 and 400 sensors that provide live data to AI systems, which make tactical recommendations that teams can use to give them an edge. Amazon Web Services (AWS) provides F1 teams with a Cloud-storage-and-computing platform that enables them to access historical data that can be used in predictive models.

Given the high incidence of accidents on our roads, the technology used by Discovery could be rolled out to other types of vehicles, especially trucks and minibus taxis, which have a high risk of being involved in accidents. Similar to Discovery's rewards system, an incentives programme could be rolled out for the drivers of these vehicles.

Studying the use of technology in sports in Africa

In 2019, I attended the International University Sports Federation (FISU) Winter Games in Siberia, Russia. I attended as a representative of the African universities on the FISU Academic Advisory Board, and I was invited to give a presentation on innovation around sports. This gave me occasion to reflect on where we are in Africa in terms of the use of technology in sports, as well as when it comes to studying tech in sports.

One of the industries not yet fully developed in Africa is sports technology. Little research is conducted around sports technology in Africa, and much of the equipment we use is imported from overseas. A challenge for industries and universities is how South Africa can develop a manufacturing industry around sports technology. We already have Sports Science centres in South Africa that are used widely, especially in competitive sports, such as swimming, athletics and others. The question is whether we are innovative in the development and usage of technology.

An example of a research project involving the use of AI in sports is the one at UJ that studies the relationship between music and motivation. This involves measuring how different types of music impact the motivation levels of groups of people while they are exercising. This raises questions, such as how AI can be used to maximise physical performance.

Another example of beneficial research involving AI and physical exercise is that done by a former Master's student of mine, Abdul-Khaliq Mohamed, who used AI to map the relationship between the movement of body parts and the activities of the brain. This is useful in treating patients who have suffered brain damage (usually caused by strokes), as it can be used to prescribe particular physical exercises involving different body parts that can target and heal certain areas of the brain.

The impact of 4IR technologies on how we watch and experience sports

In 2019, it was reported that plans were underway to make the Tokyo Olympics, set for 2020 but postponed to 2021 due to the COVID-19 pandemic, the most technologically enhanced one yet for spectators. This would be made possible through a partnership between Olympics organisers and American tech company, Intel. The plans included the 3D tracking of athletes, large-scale use of VR, AR and Mixed Reality (MR – which includes the use of VR and AR). Spectators would be able to use their smart devices to access detailed athlete profiles.

In the rapidly developing technology of the gaming space, the use of AR/VR allows users with little or no sporting ability to simulate scenarios in which they are participating in sports.

The emergence of 5G, coupled with customised apps, will enable enhanced experiences of big-ticket events, such as the Olympics – not only for those spectators who are physically present but also for mega audiences that have expectations of live streaming, interactive sessions and real-time updates. The utilisation of multiple technologies to enhance the experience of viewers or listeners will, however, require concomitant infrastructure and product development.

Elevating sports in the 4IR

AI is, therefore, disrupting our notions of sport. Rapid advances in technology are changing how we play, watch and experience it, while it is leading to new innovations in sport. The 4IR's effects on sport may be invisible to the naked eye and to those not directly involved in the industry, but the 4IR is a significant contributor to how the sports industry is changing, as well as to how athletes are aspiring to higher levels of performance. Big data and complex algorithms provide a competitive edge. Furthermore, brands are leveraging the alliance between sports, marketing and technology to promote a wide array of goods, from shoes to clothing to devices.

Companies are also using technology to customise and automate production processes. For example, Adidas has the capacity to customise shoes through automation. Companies are also using analytics and AI to identify the best materials to use in the manufacture of sportswear that is customised for the specific needs of long-distance running, for instance. In horse racing, AI-based machines are used to predict winners using data of past races.

The 4IR thus takes sport to a sophisticated new level that may need to be accompanied by suitable regulation. The danger of technological enhancement, if used unwisely, of course, is that it could make sporting excellence a dead-end pursuit. At the end of the day, we must remember that at the heart of any sport is fun. As American author Joseph Chilton Pearce put it, 'Play is the only way the highest intelligence of humankind can unfold.'

CHAPTER 26

Memory

In 2019, I visited the National Museum of African American History and Culture in Washington, DC. The museum covers all aspects of the history of African Americans, including, for example, the evolution of music such as jazz, as well as rock and roll, which African Americans introduced to the West. However, what left an indelible mark on my trip to this museum was the issue of slavery. The idea of storing memory of significant matters such as slavery is something that we should adopt very seriously. The question, however, is how to do this with the tools that we have at our disposal and to ensure that multiple perspectives are captured. This is much like creating a digital heritage. In earlier chapters, we touched on the importance of archiving our histories. This is particularly important to not only learn from the past in order to progress, but also to ensure that we do not repeat the mistakes of the past.

Apartheid history and the colonial history of South Africa have been called into question over the last 25 years. Much of this history has reflected the bias of the writers. As a result, there are generations of South Africans who have a warped understanding of the history of our country. While this has been changed at schools and universities, more needs to be done to correct history. And whereas the TRC compelled people to delve into their memories and to reconstruct events that were largely hidden from the public, it also brought into sharp focus the need to store these memories in archives that can be accessed by all.

An important question we need to consider is: what do countries with fractured histories do to rectify the collective understanding of the past, and to reconstruct those narratives using truth as opposed to warped versions of the truth?

In countries such as Rwanda and South Africa, memory, history and the reshaping of narratives have dominated discussions precisely due to the nature of the conflict in these countries and the need to understand the linkage between memory and history. Both nations are works in progress when it comes to restoring the collective memory. It must be pointed out that race, class and gender impact on who dominates in historical narratives, as well as on who is excluded or marginalised in telling these stories. It was American author and activist James Baldwin who once said, 'People are trapped in history and history is trapped in them.'

However, human beings are not the only ones with memory. Even our planet has memory. If we pollute our planet and increase greenhouse gases, it never forgets. Thousands of years from today, it will remember what we have done to it. Geological memory allows us to use techniques such as carbon dating to uncover and understand what happened a million years ago. Because of geological memory, we now know that Africa is the cradle of humankind, where human beings originated. These dating methods enable geologists and historians to pinpoint the age of objects, locate events in history, and provide a window into what kind of life prevailed when, where and how.

Preserving memory, whether geological or human, is, therefore, a pivotal project for us to undertake in the 4IR. But what can we learn, specifically, about how the memory of slavery has been constructed at museums such as the National Museum of African American History and Culture? It would be useful, in this respect, to look at the causes of slavery more broadly.

Learning from the history of slavery

Slavery essentially happened because of greed, a trait that makes humans evil. On justifying slavery, slave owner William Snelgrave said in 1734: 'Tho' to traffic in human creatures may at first appear barbarous, the advantage of it far outweighs the inconvenience.' Similarly, William Cowper said: 'I admit I am sickened at the purchase of slaves … but I must be mum, for how could we do without sugar or rum?'

Slavery also happened because humans are hypocrites. The United States was founded at a time when slavery was taking place at full tilt. The founding fathers of the United States were slave owners themselves. George Washington and Thomas Jefferson owned 123 slaves and 600 slaves, respectively. Jefferson even enslaved his own children. However, despite the fact that they owned slaves, this did not stop them from writing that 'all men are created equal' in the Declaration of Independence. These founding fathers did not appreciate the universality of truth. Humans can be honest and dishonest at the same time; they can tell the truth and lie in the same breath. These are very real moral dilemmas. Humans are creatures of contradiction and are hopelessly irrational. Great men and women are those who, every day, attempt to resolve their contradictions and irrationality.

It can also be argued that one cause of slavery, notwithstanding the hypocrisy and greed of slave owners, is information asymmetry, which as we have seen in previous chapters is an economic term that means that one party in a transaction has greater knowledge, or more information, than the other party. Whenever information asymmetry exists, we see exploitation. In the case of slavery, the enslavers had better technology than the enslaved.

Information asymmetry explains, in part, many historical events. It contributed to the colonisation of India by Britain. It contributed to the Israeli–Palestinian conflict and much of what happens in the Middle East. It contributed to the Berlin Conference that divided the African continent into colonies of Europe. The territorial boundaries resulting from this conference have existed up until today, yet they make no geographical sense.

'For years, Africa has been the footstool of colonialism and imperialism, exploitation and degradation ... Those days are gone and gone forever, and now I, an African, stand before this august Assembly of the United Nations and speak with a voice of peace and freedom, proclaiming to the world the dawn of a new era ... There are now 22 of us and there are yet more to come.' These were the words of Kwame Nkrumah, president of Ghana, in 1960, the year sixteen African countries joined the UN to take their seats at the table previously reserved for the coloniser. Let us collapse these colonial borders!

Because of slavery, nations such as Portugal, Britain, France, the Netherlands and Denmark became so wealthy that they still benefit from the results of their plundering today. Portugal took 5.8 million African people into slavery. Similarly, Britain took 3.3 million people; France took 1.4 million people; the Netherlands took 555 300 people; and Denmark took 85 000 African people into slavery. When those slaves or their descendants were finally 'freed', they received no compensation – whereas slave owners received compensation of the equivalent of £17 billion. Slaves built the White House, yet only one of their descendants has ever occupied it as president of the United States: Barack Obama, who is descended, on his mother's side, from John Punch, a servant who was forced into slavery in 1640 and who some historians consider the first African enslaved in the colonies.[1]

Slavery also happened because of selfishness, which goes against the theory of utilitarianism. As we saw earlier in the book, utilitarianism is that which brings the greatest amount of happiness to the greatest number of people. The problem with this is that in justifying over-using some resources for the greater benefit of all, humans often destroy other important assets. For example, while maximising profits, the mining companies in South Africa exploited the workers and left dangerous radioactive yellow mountains like those located next to Soweto. These mountains are so dangerous that we still have not quantified how many people they kill by way of disease every year. For communities living in close proximity to the mines, the jagged edge of the knife has been the health-related side effects that amount to the worst public health disaster that exploited the differences between ethnic groups in South Africa's history.

The enslavers' notion of tribalism sought to find differences where none existed. It can be argued that the enslavers in Africa used their conceptualisation of these differences in order to facilitate slavery. For example, in Rwanda, the Belgians were instrumental in 'creating' the Tutsi ethnicity, by classifying people according to the measurements of different parts of their bodies, as well as by counting the number of cattle they owned. In the 1994 Rwandan genocide, the Hutus killed a million Tutsis because they considered them different. But what is the difference between the Hutus and Tutsis? Do they not speak the same language?

Slavery also happened because evil is more aggressive than good is. In her classic book, *Eichmann in Jerusalem: A Report on the Banality of Evil*, philosopher Hannah Arendt studied how such an 'ordinary' man, Adolf Eichmann, ended up as a Nazi who killed hundreds of thousands of people. To make sense of this, she coined the term 'the banality of evil', which theorises that evilness can be passive, not necessarily only committed by evil people but also by bureaucrats dutifully obeying their orders. I think Arendt was wrong; I think evil is very aggressive and she could, therefore, have appropriately used the term 'the aggression of evil' to describe Eichmann. Eichmann voluntarily joined the Nazis and aggressively did everything to ascend through the Nazi hierarchy. Arendt's apparent passivity of evil is nothing but a deception to pacify those with good intentions. This form of 'aggressive evil' that I am describing has been repeated across the centuries in different parts of the world with devastating consequences. Apartheid was a series of multiple acts of aggressive evil and was, in 1973, referred to as a crime against humanity by the United Nations General Assembly.

Celebrating forgotten heroes

Creating a collective memory is also about giving space to those who have been forgotten as a result of biased historical storytelling. One such example is Dr Katherine Johnson, one of the first African Americans to enrol in the mathematics programme at West Virginia University, and an unsung hero of America's first moon landing. Dr Johnson passed away in February 2020.

During her more than three-decades-long career at NASA, she earned a reputation for mastering complex manual calculations, combining her mathematical talent with computer skills to solve problems of an astrophysics nature. Her mathematical skills helped put US astronaut John Glenn into orbit around the Earth in 1962 and helped calculate the trajectory for the 1969 *Apollo 11* flight to the moon. Yet, Dr Johnson's history was steeped in the history of US segregation. Even during the space race between the US and the Soviet Union, in which she played a pivotal role, Johnson and her African American colleagues worked in separate facilities to white workers and used different toilets and dining areas.

It is heroes such as these that we need to celebrate in our collective memory as we reconstruct history in the time of the 4IR.

Using AI to help us avoid repeating history

Karl Marx once said: 'History repeats itself first as a tragedy and second as a farce.' We should not take lightly the idea that history repeats itself. After Nelson Mandela made his inaugural speech as president of South Africa, in which he stated, 'Never, never and never again shall it be that this beautiful land will again experience the oppression of one by another,' the Rwandan genocide continued to unfold.

History also tends to repeat itself in times of dire economic downturns. For example, in South Africa, xenophobic attacks took place against foreign nationals considered to be 'outsiders'; 'those who were taking South Africans' jobs'. Therefore, we need to be vigilant in our defence of human dignity and liberty. How do we ensure that we retain our memories of the past in order to avoid the pitfalls going into the future?

One solution is to use AI to help us remember. As we have seen in previous chapters, AI gives machines the ability to mimic or improve on human behaviour. Furthermore, it can be enhanced with the ability to recognise patterns and suggest solutions.

We already rely on technology to augment our memory. For example, search engines such as Google make it easy for us to access information about the past, meaning that we don't need to store this knowledge in our own memories.

Smartphones circumvent the need for us to remember telephone numbers, while Facebook even reduces the need to remember the names of people in images, because it suggests names to tag them with. This capacity has been greatly enhanced by advances in AI predicated on deep learning, which as we saw in earlier chapters involves large neural networks with multiple layers. In the medical field, the storing of thousands of images or data sets enables enhanced surgical procedures that make use of robots. The programming or sophistication of the robots is rooted in machine-learning algorithms that guide them in using mimicry and repetition in order to derive precision.

In education, many apps encourage and trigger memory retention in students. Students with language barriers and other impediments can rely on a variety of technologies to enhance and deepen the learning experience. For example, a student for whom English was a second-language once explained to me that having access to a podcast of a Physics lecture was critical for them to understand and learn. Exposure to another language in your first year can be daunting, so the podcast allowed such students to listen, pause and assimilate as many times as required. This breaks down not only barriers to learning but also language barriers that would otherwise impede a student's progress.

What, then, is to be done in order for us to avoid repeating history? As the poet Maya Angelou once said, 'History, despite its wrenching pain, cannot be unlived, but if faced with courage, need not be lived again.'

Firstly, we need to resolve the information asymmetry that exists in society. We are living in an era in which advances in intelligent technologies will change human identity, our environment, our politics, our economy and our society. The 4IR will widen income gaps, increase inequality, introduce new forms of slavery and thus sharpen the contradictions – and, in Marxist thinking, sharpening the contradictions naturally leads to violent revolution. Countries that understand and harness advances in the 4IR will hold the power; as the president of Russia, Vladimir Putin said, 'Those who master AI will control the world.' In fact, nations could become so powerful that, if this control is not checked, it can lead to a new form of slavery.

Secondly, we need to regulate greed. The 2008 financial crash that impoverished many people and many nations came about because of greed. Economic growth should not come about as a result of a desire to maximise profit at the expense of human dignity.

Thirdly, we need to banish hatred in all its manifestations. We need to educate people in order to create and nurture memories so that we do not repeat mistakes. As former US president JF Kennedy once said: 'Forgive your enemies but never forget their names.' To remember their names, we need to memorise them.

It is also important to understand how technologies of the 4IR work in replicating human memory and what we can learn from that. Perhaps this is the best way to ensure that history does not repeat itself. Algorithms mimic human thinking by picking up on and replicating patterns. However, human memory is far more complicated than you might think. For example, if you saw a man driving a car one day and then saw an older woman driving the same car the next day, you might have subconsciously made some inferences. You may have inferred that the man and the older woman were from the same family; perhaps that they shared a car; perhaps she was just a friend borrowing the car, or perhaps she was his mother. This is what we call 'episodic memory' – a type of memory that machines find very difficult to replicate. In its current state, technologies of the 4IR cannot match human episodic memory. UK AI company DeepMind Technologies is currently conducting research on whether neuroscience can inspire more sophisticated AI. In fact, DeepMind has devised a new kind of computer that has a working memory. The researchers show that the computer, which consists of a large neural network connected to a unique form of memory, can perform relatively complex tasks by figuring out for itself what information to hold in its memory. It can, for example, navigate London's complex Underground rail system. According to a paper by its developers, 'Like a conventional computer, it can use its memory to represent and manipulate complex data structures but, like a neural network, it can learn to do so from data.'[2]

While this notion is still in its infancy, AI has been able to replicate some aspects of human cognitive abilities. For example, techniques such as Neural Turing Machines (NTM) have made significant progress in setting up the foundation for building human-like memory structures in deep-learning systems. This, of course, needs to be interwoven with the creation of ethical AI, one in which it is able to tell which decisions are bad and should not be repeated and which decisions can be replicated. This is particularly important as AI automates more government-related functions.

Conclusion

Kwame Nkrumah, the former Ghanaian president and a visionary with a dream of what we in Africa can become, once said, 'We shall accumulate machinery and establish steelworks, iron foundries and factories; we shall link the various states of our continent with communications; we shall astound the world with our hydroelectric power; we shall drain marshes and swamps, clear infested areas, feed the undernourished, and rid our people of parasites and disease. It is within the possibility of science and technology to make even the Sahara bloom into a vast field with verdant vegetation for agricultural and industrial developments.'

This speech was made more than 50 years ago. Embedded in the vision of Nkrumah was that Africa needs to rise and that, together, the continent could prove to be a force. Innovation, it is believed, largely emerges elsewhere, with Africa as a receiving agent. In other words, we are – realistically speaking – not innovating in the same way as the US, China and other countries. Of course, there are varying schools of thought on the 4IR. On the one hand, many think it is the key to our fortunes. On the other hand, there is a warranted fear that it will widen our disparities and worsen many of the challenges we face as a continent.

Yet, as we track the use of 4IR technologies in economies, across industries and in society, there are pockets of opportunity for the African continent that we simply cannot afford to miss. Here we find ourselves at a crossroads, much like the poet Robert Frost, who wrote, 'Two roads diverged in a yellow wood, And sorry I could not travel both.' One road could retreat from the realities of the 4IR, with no prospect of development, but it would be in our own best interests to remember that the 4IR is an era in which intelligent technologies permeate all aspects of our lives, be they in the economy, society or politics.

This book has outlined how the 4IR is fundamentally changing every facet of our lives. It has traced how economies are transforming, how we can begin to answer questions around financial inclusion, unemployment and lack-lustre growth. We have looked at how industries are fundamentally evolving and the moves we need to make to rise to the challenges the 4IR presents. Our fortunes are certainly no longer located in the extractive industries; we have to push for innovation in other industries, so as to not only curb de-industrialisation but also to remain relevant and competitive amid a rapidly changing context. Our societies also need to adapt, and we must now focus on the creation of our own techniques and our own systems – those that are unique to the African context.

While there are many instances of the developed world making strides in this era, the African continent has a lot to boast about. Yet, there is room to move, and to do so with agility. In many ways, we may have seen the last three industrial revolutions pass us by, but we cannot afford to be passive observers of this one. The continent has the potential to be a leader and contributor to the developments of the 4IR. We are tasked to ensure that Africa provides at least some answers to the challenges raised. As President Cyril Ramaphosa put it, 'We were left behind by the first industrial revolution, the second and so forth, but the fourth one is not going to leave us behind – we are going to get ahead of that fourth industrial revolution.'

Notes

Chapter 1: Understanding the fourth industrial revolution

1 David Mills (14 January 2020), 'Welcoming the Fourth Industrial Revolution', *Forbes*, https://www.forbes.com/sites/ricoheurope/2020/01/14/welcoming-the-fourth-industrial-evolution/#308e7619473a.
2 Wikipedia, 'Reformation', https://en.wikipedia.org/wiki/Reformation.
3 South African History Online, 'The Industrial Revolution in Britain and Southern Africa from 1860', https://www.sahistory.org.za/article/grade-8-term-1-industrial-revolution-britain-and-southern-africa-1860-0.
4 Ibid.
5 Dirk de Vos (24 May 2017), 'Analysis: South Africa is Stuck in a Failing Second Industrial Revolution – Let's Move Beyond It', *Daily Maverick*, https://www.dailymaverick.co.za/article/2017-05-24-analysis-south-africa-is-stuck-in-a-failing-second-industrial-revolution-lets-move-beyond-it/.
6 Khadija Sharife and Patrick Bond (2011), 'Above and Beyond South Africa's Minerals-Energy Complex', in D. Pillay, J. Daniels, P. Naidoo and R. Southall (eds), *New South African Review 2* (Johannesburg: Wits University Press), http://www.ee.co.za/wp-content/uploads/legacy/Sharife-Bond-MEC-in-New-SA-Review-2.pdf.
7 Chris Woodford (29 June 2019), 'Transistors', Explainthatstuff.com, https://www.explainthatstuff.com/howtransistorswork.html.
8 Ibid.
9 DeepAI, 'What is a Hidden Layer?', https://deepai.org/machine-learning-glossary-and-terms/hidden-layer-machine-learning.
10 Eugène N. Marais (2010), *Eugène Nielen Marais Collection* (retrieved 2 July 2020), from https://ujdigispace.uj.ac.za/handle/10210/3362.
11 Johann L. Marais (2005), *Eugene N. Marais se Joernalistieke en Bellettristiese Prosa* (retrieved 2 July 2020), from https://journals.co.za/content/stilet/17/1/ejc109784, and Martin S. Olivier (2015). '"Die Siel van die Mier": Reflections on the Battle for "Scholarly" Intelligence', *The Journal for Transdisciplinary Research in Southern Africa*, 11(2), (retrieved 2 July 2020), from https://repository.up.ac.za/handle/2263/51112.
12 Saibal Dasgupta (19 April 2019), 'China's Political System Helps Advance Its Artificial Intelligence', Voice of America, https://www.voanews.com/east-asia-pacific/chinas-political-system-helps-advance-its-artificial-intelligence.
13 Fabian Westerheide (14 January 2020), 'China – The First Artificial Intelligence Superpower', *Forbes*, https://www.forbes.com/sites/cognitiveworld/2020/01/14/china-artificial-intelligence-superpower/#13bf3bd02f05.
14 Future of Life Institute, 'AI Policy – Kenya', https://futureoflife.org/ai-policy-kenya/.

Chapter 2: Automation

1 Kenneth Rogoff (2 October 2012), 'The Impact of Technology on Employment', World

Economic Forum, https://www.weforum.org/agenda/2012/10/king-ludd-is-still-dead/.

2 Lenyaro Sello (3 July 2019), '4IR and Africa: Promise or Peril?', Investec, https://www.investec.com/en_za/focus/innovation/4ir-and-africa-a-digital-divide.html.

3 Bruce Lee (2011 [1975]), *Tao of Jeet Kune Do* (Valencia, California: Black Belt Communications), https://www.amazon.com/Tao-Jeet-Kune-Bruce-Lee-ebook/dp/B0052FYPJK/ref=tmm_kin_swatch_0?_encoding=UTF8&qid=&sr=.

4 Bob Violino (1 April 2019), 'How to Thrive in a "Business 4.0" World', ZD Net, https://www.zdnet.com/article/how-to-thrive-in-a-business-4-0-world/.

5 International Data Corporation (4 September 2019), 'Worldwide Spending on Artificial Intelligence Systems Will be Nearly $98 Billion in 2023, According to New IDC Spending Guide', https://www.idc.com/getdoc.jsp?containerId=prUS45481219.

Chapter 3: South Africa's 4IR strategy

1 4IRSA, 'South Africa is a Mix Bag on its Readiness for 4IR', https://4irsa.org/south-africa-4-0/south-africa-is-a-mix-bag-on-its-readiness-for-4ir/#.

Chapter 4: Safety of structures

1 Tshilidzi Marwala (2012), *Condition Monitoring Using Computational Intelligence Methods* (London: Springer).

2 Tshilidzi Marwala (2014), *Artificial Intelligence Techniques for Rational Decision Making* (London: Springer).

Chapter 6: Mining

1 Gareth van Zyl (18 March 2019), 'Ivo Vegter: Here's Why Mining Still Matters if SA is to Thrive', BizNews, https://www.biznews.com/thought-leaders/2019/03/18/ivo-vegter-why-mining-matters-sa.

2 Yuval N. Harari (2018), *21 Lessons for the 21st Century* (London: Jonathan Cape).

Chapter 8: Data privacy

1 Daniel Kahneman (2011), *Thinking, Fast and Slow* (New York: Farrar, Straus and Giroux).

2 Eames Yates (25 March 2017), 'What Happens to Your Brain when You Get a Like on Instagram', Business Insider, https://www.businessinsider.com/what-happens-to-your-brain-like-instagram-dopamine-2017-3?IR=T.

3 Julian Morgans (19 May 2017), 'Your Addiction to Social Media Is No Accident', Vice, https://www.vice.com/en_us/article/vv5jkb/the-secret-ways-social-media-is-built-for-addiction.

4 Businessballs, 'Nudge Theory', https://www.businessballs.com/improving-workplace-performance/nudge-theory/.

Chapter 9: Digital heritage

1 Beth Daley (15 July 2019), 'Importance of Digitising Cultural Heritage Highlighted in "Heritage at Risk" Exhibition', Europeana, https://pro.europeana.eu/post/importance-of-digitising-cultural-heritage-highlighted-in-heritage-at-risk-exhibition.

2 Ibid.

3 Vincent Matinde (20 January 2017), 'Africa's First Programming Language to Teach Kids Code', IDG Connect, https://www.idgconnect.com/idgconnect/interviews/1001457/africas-programming-language-teach-kids-code.

Chapter 10: Cybersecurity

1 Oliver Tambo (1964), 'Nelson Mandela', South African History Online, https://www.sahistory.org.za/archive/nelson-mandela-oliver-tambo.
2 Carin Smith (29 April 2019), 'Major Spike in SA Cyber Attacks, over 10 000 Attempts a Day – Security Company', fin24, https://www.fin24.com/Companies/ICT/major-spike-in-sa-cyber-attacks-over-10-000-attempts-a-day-security-company-20190429.
3 Brian Joss (12 June 2019), 'SA Lags behind on the Cybersecurity Front', IOL, https://www.iol.co.za/personal-finance/sa-lags-behind-on-the-cybersecurity-front-25801597.
4 Riaan Grobler (25 October 2019), 'City of Joburg Shuts Down All Systems after Cyber Attack Demanding Bitcoin Ransom', news24, https://www.news24.com/news24/southafrica/news/city-of-joburg-shuts-down-all-systems-after-cyber-attack-demanding-bitcoin-ransom-20191025.
5 Ben Evansky (5 March 2020), 'US, UK and Estonia Call out Russia over Cyber Attacks against Georgia in UN Security Council First', Fox News, https://www.foxnews.com/world/us-uk-estonia-call-out-russia-cyber-attacks-against-georgia.
6 Douglas Bonderud (14 June 2018), 'Enterprise Cloud Security: Is Blockchain Technology the Missing Link?', SecurityIntelligence, https://securityintelligence.com/enterprise-cloud-security-is-blockchain-technology-the-missing-link/.

Chapter 11: Social networks

1 Statista (April 2020), 'Most Popular Social Networks Worldwide as of April 2020, Ranked by Number of Active Users', https://www.statista.com/statistics/272014/global-social-networks-ranked-by-number-of-users/.
2 Sam Byford (30 November 2018), 'How China's Bytedance Became the World's Most Valuable Startup', The Verge, https://www.theverge.com/2018/11/30/18107732/bytedance-valuation-tiktok-china-startup.
3 Ibid.
4 Dianna Christe (9 April 2019), 'Gen Z Prefers Instagram when Hearing from Brands, Piper Jaffray Says', Marketing Dive, https://www.marketingdive.com/news/gen-z-prefers-instagram-when-hearing-from-brands-piper-jaffray-says/552296/.
5 Jessica Davis (31 August 2018), 'How Artificial Intelligence Models are Taking over Your Instagram Feed', Bazaar, https://www.harpersbazaar.com/uk/fashion/fashion-news/a22722480/how-artificial-intelligence-models-are-taking-over-your-instagram-feed/.
6 Ibid.
7 Ibid.

Chapter 12: 5G technology

1 Simnikiwe Mzekandaba (28 October 2019), 'Release of 5G Spectrum Won't be Delayed, Minister Promises', ITWeb, https://www.itweb.co.za/content/Pero3qZg1oRvQb6m.
2 Ibid.
3 Arjun Kharpal (25 February 2020), 'Op-Ed: America Has Limited Options on 5G to Fend off China's Huawei Challenge', CNBC, https://www.cnbc.com/2020/02/25/america-has-limited-options-on-5g-to-fend-off-chinas-huawei-challenge.html.
4 TechCentral (2 February 2020), 'Submissions Pour in on Icasa Spectrum Licensing Plan', https://techcentral.co.za/submissions-pour-in-on-icasa-spectrum-licensing-plan/95611/.
5 Ibid.
6 TimesLIVE (2 February 2020), 'Deadline Extended for Vodacom and MTN Deal on Cheaper Data Prices', https://www.timeslive.co.za/news/south-africa/2020-02-02-deadline-extended-for-vodacom-and-mtn-deal-on-cheaper-data-prices/.

7 Ibid.
8 Dan Meyer (20 March 2020), 'Data Price Cuts: MTN Join Vodacom in Massive Price Slash', The South African, https://www.thesouthafrican.com/technology/data-price-cuts-mtn-vodacom-competition-commission-2020/.
9 Joanne Carew (15 November 2019), 'There Is No Fourth Industrial Revolution without 5G', ITWeb, https://www.itweb.co.za/content/VgZey7JABzevdjX9.
10 Ben Mankidimanda (10 April 2020), 'The Evolution of Wireless Generations', Special World for Life Livings, https://specialworldforlife.blog/2020/04/10/the-evolution-of-wireless-generations/.
11 Ibid.
12 Ibid.
13 Ibid.
14 McKinsey Global Institute (20 February 2020), 'Connected World: An Evolution in Connectivity Beyond the 5G Revolution', McKinsey & Company, https://www.mckinsey.com/industries/technology-media-and-telecommunications/our-insights/connected-world-an-evolution-in-connectivity-beyond-the-5g-revolution.
15 Ibid.
16 Ibid.
17 Liquid Telecom (20 January 2020), 'Liquid Telecom to Launch First 5G Wholesale Roaming Network Service in South Africa', https://www.liquidtelecom.com/about-us/news/Liquid%20Telecom%20to%20launch%20first%205G%20wholesale%20roaming%20network%20service%20in%20South%20Africa.

Chapter 13: Economics

1 Siphelele Dludla (21 May 2020), 'More Capital for Households as SA Reserve Bank Relaxes Monetary Policy', IOL, https://www.iol.co.za/business-report/economy/more-capital-for-households-as-sa-reserve-bank-relaxes-monetary-policy-48328396.
2 Irving Wladawsky-Berger (16 November 2018), 'The Impact of Artificial Intelligence on the World Economy', *The Wall Street Journal*, https://blogs.wsj.com/cio/2018/11/16/the-impact-of-artificial-intelligence-on-the-world-economy/.
3 Ibid.
4 Ibid.
5 Ibid.
6 Accenture, 'Fuel for Growth', https://www.accenture.com/sk-en/insight-artificial-intelligence-future-growth.
7 Adam Shaw (2 September 2017), 'Why Economic Forecasting Has Always Been a Flawed Science', *The Guardian*, https://www.theguardian.com/money/2017/sep/02/economic-forecasting-flawed-science-data.
8 Asanda Fotoyi (2017), 'Revisions to South Africa's Gross Domestic Product', Trade & Industrial Policy Strategies, https://www.tips.org.za/policy-briefs/item/3399-revisions-to-south-africa-s-gross-domestic-product.

Chapter 14: Banking

1 Lesetja Kganyago, 'Foreword', in 'The Impact of the 4th Industrial Revolution on the South African Financial Services Market', Centre of Excellence in Financial Services, https://www.genesis-analytics.com/uploads/downloads/COEFS-Theimpactofthefourthindustrialrevolutiononfinancialservicesin SouthAfrica-final-1-FR.pdf.
2 Centre of Excellence in Financial Services, 'The Impact of the 4th Industrial Revolution on the South African Financial Services Market', https://www.genesis-analytics.com/uploads/

downloads/COEFS-TheimpactofthefourthindustrialrevolutiononfinancialservicesinSouth Africa-final-1-FR.pdf.
3 Ibid.
4 Bloomberg (22 October 2019), 'Half the World's Banks Are Too Weak to Survive a Downturn, McKinsey Says', *Business Maverick*, https://www.dailymaverick.co.za/article/2019-10-22-half-the-worlds-banks-are-too-weak-to-survive-a-downturn-mckinsey-says/.
5 Ibid.
6 Business Tech (13 November 2019), 'Discovery vs TymeBank vs Bank Zero – What South Africans Think of the New Banks', https://businesstech.co.za/news/banking/352515/discovery-vs-tyme-bank-vs-bank-zero-what-south-africans-think-of-the-new-banks/.
7 Tshilidzi Marwala (2013), *Economic Modeling Using Artificial Intelligence Methods* (London: Springer).
8 Robert Williams (24 October 2019), 'Apple Pay Surpasses Starbucks as Most Popular Mobile Payment App', Mobile Marketer, https://www.mobilemarketer.com/news/apple-pay-surpasses-starbucks-as-most-popular-mobile-payment-app/565726/.
9 Ibid.
10 Samsung (11 March 2019), 'Samsung Pay Expands Reach to Millions of South Africans and Soon to Launch with More Banks', https://news.samsung.com/za/samsung-pay-expands-reach-to-millions-of-south-africans-and-soon-with-even-more-banks.
11 Lyal White and Liezl Rees (1 October 2019), 'The Digital Revolution Can Unlock Inclusive Growth and More Jobs in Africa', BizCommunity, https://www.bizcommunity.com/Article/196/831/196186.html.
12 Alex Liu (1 September 2019), 'Africa's Future is Innovation Rather than Industrialization', World Economic Forum, https://www.weforum.org/agenda/2019/09/africa-innovation-rather-than-industrialization/.
13 Linus Unah (6 November 2019), 'Andela Layoffs Mark Shift in Youth Talent Policy', African Business, https://africanbusinessmagazine.com/sectors/technology/andela-layoffs-mark-shift-in-youth-talent-policy/.
14 Greg Chen (31 October 2019), 'Addressing Data Inequality is Key to Job Creation in South Africa', Tech Financials, https://techfinancials.co.za/2019/10/31/addressing-data-inequality-is-key-to-job-creation-in-south-africa/.
15 Damian Radcliffe (16 October 2018), 'Mobile in Sub-Saharan Africa: Can World's Fastest-Growing Mobile Region Keep It Up?', ZD Net, https://www.zdnet.com/article/mobile-in-sub-saharan-africa-can-worlds-fastest-growing-mobile-region-keep-it-up/.

Chapter 15: Taxation

1 Business Tech (3 February 2020), 'New SARS Technology to Go after Tax-Dodging South Africans', https://businesstech.co.za/news/technology/370178/new-sars-technology-to-go-after-tax-dodging-south-africans/.
2 Kabous le Roux (12 February 2020), 'Artificial Intelligence Running on Supercomputers are Coming for Tax Dodgers', Cape Talk, http://www.capetalk.co.za/articles/374754/sars-is-going-high-tech-artificial-intelligence-running-on-supercomputers-are-coming-for-tax-dodgers.
3 Business Wire (8 December 2014), 'CPA.com Study Takes Pulse of CPA of the Future', https://www.businesswire.com/news/home/20141208005814/en/CPA.com-Study-Takes-Pulse-of.
4 Ibid.
5 Rose Leadem (20 February 2017), 'Bill Gates Believes Robots that Steal Jobs Should Pay Taxes', Entrepreneur South Africa, https://www.entrepreneur.com/article/289491.

6 Ibid.
7 Tshilidzi Marwala and Evan Hurwitz (2017), *Artificial Intelligence and Economic Theory: Skynet in the Market* (London: Springer).

Chapter 16: Market efficiency

1 Daniel Silke (9 December 2016), 'The Nenegate Legacy – What Doesn't Kill You Makes You Stronger', BizNews, https://www.biznews.com/leadership/2016/12/09/nenegate-legacy-daniel-silke.
2 Mike Cohen and Colleen Goko (10 October 2018), 'Another Day Another Finance Minister – But Market Favours Mboweni', BizNews, https://www.biznews.com/sa-investing/2018/10/10/finance-minister-market-favours-mboweni.
3 Tshilidzi Marwala (2 November 2018), 'Are South African Markets Efficient?', https://www.dailymaverick.co.za/opinionista/2018-11-02-are-south-african-markets-efficient/#gsc.tab=0.
4 Johannesburg Stock Exchange (2013), 'Market Announcements', https://www.jse.co.za/services/market-data/market-announcements.
5 Dave Mohr and Izak Odendaal (12 December 2017), 'Lessons Steinhoff Taught Us about Investing', CNBC Africa, https://www.cnbcafrica.com/insights/steinhoff/2017/12/12/steinhoff-lessons/.
6 Adair Turner (24 January 2012), '"Efficient Markets Hypothesis" Inefficient', *Financial Times*, https://www.ft.com/content/cb7e1b6e-46bc-11e1-bc5f-00144feabdc0.
7 Karl Kaufman (30 September 2018), '"The Market Is Always Wrong": In Defense of Inefficiency', *Forbes*, https://www.forbes.com/sites/karlkaufman/2018/09/30/the-market-is-always-wrong-in-defense-of-inefficiency/#259f32a44da1.
8 Daniel Kahneman (2011). Thinking, Fast and Slow (New York: Farrar, Straus and Giroux).
9 Mike Thomas (16 March 2019), 'How AI Trading Technology is Making Stock Market Investors Smarter', Built In, https://builtin.com/artificial-intelligence/ai-trading-stock-market-tech.
10 Ibid.
11 Ibid.
12 MarketInsite (7 November 2019), 'For the First Time, Nasdaq is Using Artificial Intelligence to Surveil US Stock Market', Nasdaq, https://www.nasdaq.com/articles/for-the-first-time-nasdaq-is-using-artificial-intelligence-to-surveil-u.s.-stock-market.
13 Federal Reserve Bank of San Francisco (January 2005), 'Education: Please Explain How Financial Markets May Affect Economic Performance', https://www.frbsf.org/education/publications/doctor-econ/2005/january/financial-markets-economic-performance/.
14 Kai-Fu Lee (2018), *AI Superpowers: China, Silicon Valley, and the New World Order* (New York: Houghton Mifflin Harcourt).

Chapter 17: Trade

1 Mzala Nxumalo (1985), 'Cooking the Rice Inside the Pot: A Historical Call in Our Times', https://docs.google.com/viewer?a=v&pid=sites&srcid=ZGVmYXVsdGRvbWFpbnxjb21tdW5pc3R1bml2ZXJzaXR5YW5uZXhlfGd4OjYxYTljYmM0NDI2YTYxMTc.
2 Carin Smith (5 February 2019), 'Mantashe to Investors: Don't Listen to Doomsayers, Invest in SA', fin24, https://m.fin24.com/Special-Reports/Mining-Indaba/mantashe-to-investors-dont-listen-to-doomsayers-invest-in-sa-20190205.
3 Deloitte, 'Africa Industry 4.0 Report', https://www2.deloitte.com/content/dam/Deloitte/za/Documents/manufacturing/za-Africa-industry-4.0-report-April14.pdf.
4 Ibid.

5 Ibid.
6 Business Tech (7 March 2020), 'Moody's Cuts South Africa's 2020 Growth Prospects', https://businesstech.co.za/news/finance/379781/moodys-cuts-south-africas-2020-growth-prospects/.

Chapter 18: Leadership

1 Tomas Chamorro-Premuzic, Michael Wade and Jennifer Jordan (22 January 2018), 'As AI Makes More Decisions, the Nature of Leadership Will Change', https://hbr.org/2018/01/as-ai-makes-more-decisions-the-nature-of-leadership-will-change.
2 Punit Renjen (23 January 2019), 'The 4 Types of Leader Who Will Thrive in the Fourth Industrial Revolution', World Economic Forum, https://www.weforum.org/agenda/2019/01/these-four-leadership-styles-are-key-to-success-in-the-fourth-industrial-revolution/.
3 Bo Xing and Tshilidzi Marwala (2017), 'Implications of the Fourth Industrial Age on Higher Education', https://arxiv.org/ftp/arxiv/papers/1703/1703.09643.pdf.

Chapter 19: Languages

1 Abdi Latif Dahir (27 November 2018), 'African Languages are being Left Behind When it Comes to Voice Recognition Innovation', *Quartz Africa*, https://qz.com/africa/1475763/african-languages-are-lagging-behind-when-it-comes-to-voice-recognition-innovations/.
2 Mildred Europa Taylor (2 May 2019), 'Nigerian Man Develops World's First AI Portal that Can Translate Over 2,000 African Languages', Face2face Africa, https://face2faceafrica.com/article/nigerian-man-develops-worlds-first-ai-portal-that-can-translate-over-2000-african-languages.
3 Ibid.
4 Katie Cashman (6 February 2020), 'Masakhane: Using AI to Bring African Languages into the Global Conversation', Reset, https://en.reset.org/blog/masakhane-using-ai-bring-african-languages-global-conversation-02062020.

Chapter 20: Ethics

1 Future of Life Institute, 'An Open Letter: Research Priorities for Robust and Beneficial Artificial Intelligence', https://futureoflife.org/ai-open-letter/.
2 Joseph Foley (23 March 2020), '10 Deepfake Examples that Terrified and Amused the Internet', Creative Bloq, https://www.creativebloq.com/features/deepfake-examples.
3 Tom Chatfield (16 January 2020), 'There's No Such Thing as "Ethical A.I."', OneZero, https://onezero.medium.com/theres-no-such-thing-as-ethical-a-i-38891899261d.
4 Klaus Schwab (10 February 2016), 'How Will the Fourth Industrial Revolution Affect International Security?', World Economic Forum, https://www.weforum.org/agenda/2016/02/how-will-the-fourth-industrial-revolution-affect-international-security/.
5 Jenny Anderson (4 September 2019), 'MIT Developed a Course to Teach Tweens about the Ethics of AI', *Quartz*, https://qz.com/1700325/mit-developed-a-course-to-teach-tweens-about-the-ethics-of-ai/.
6 Accenture, 'Responsible AI and Robotics: An Ethical Framework', https://www.accenture.com/gb-en/company-responsible-ai-robotics.
7 Megan J. Russell, David M. Rubin, Brian Wigdorowitz and Tshilidzi Marwala. (PCT/IB2009/006125) An Artificial Larynx.

Chapter 21: Democracy

1 Jamie Susskind (2018), *Future Politics: Living Together in a World Transformed by Tech* (Oxford: Oxford University Press), https://www.amazon.com/

Future-Politics-Living-Together-Transformed/dp/0198848927/ref=sr_1_1?dchild=1&keywords=Jamie+Susskind+%282018%29%2C+Future+Politics%3A+Living+Together+in+a+World+Transformed+by+Tech+%28Oxford%3A+Oxford+University+Press&qid=1593420786&sr=8-1.

2 Eli Pariser (22 May 2011), 'What the Internet Knows about You', CNN, https://edition.cnn.com/2011/OPINION/05/22/pariser.filter.bubble/index.html.

3 Henry Kissinger (2014), *World Order: Reflections on the Character of Nations and the Course of History* (New York: Penguin Books), https://www.amazon.com/World-Order-Henry-Kissinger-ebook/dp/B00INIXVMK/ref=tmm_kin_swatch_0?_encoding=UTF8&qid=&sr=.

4 Christina Larson (20 August 2018), 'Who Needs Democracy When You Have Data?', *MIT Technology Review*, https://www.technologyreview.com/2018/08/20/240293/who-needs-democracy-when-you-have-data/.

5 Steven Feldstein (9 January 2019), 'How Artificial Intelligence is Reshaping Repression', Carnegie Endowment for International Peace, https://carnegieendowment.org/2019/01/09/how-artificial-intelligence-is-reshaping-repression-pub-78093.

6 Steven Feldstein (22 April 2019), 'How Artificial Intelligence Systems Could Threaten Democracy', The Conversation, https://theconversation.com/how-artificial-intelligence-systems-could-threaten-democracy-109698.

7 Ibid.

Chapter 22: Movies

1 Annalee Newitz (6 September 2016), 'Movie Written by Algorithm Turns out to be Hilarious and Intense', Ars Technica, https://arstechnica.com/gaming/2016/06/an-ai-wrote-this-movie-and-its-strangely-moving/.

2 Ibid.

3 Ibid.

4 James Vincent (28 May 2019), 'Hollywood is Quietly Using AI to Help Decide Which Movies to Make', The Verge, https://www.theverge.com/2019/5/28/18637135/hollywood-ai-film-decision-script-analysis-data-machine-learning.

5 Steve Rose (16 January 2020), '"It's a War between Technology and a Donkey" – How AI is Shaking up Hollywood', *The Guardian*, https://www.theguardian.com/film/2020/jan/16/its-a-war-between-technology-and-a-donkey-how-ai-is-shaking-up-hollywood.

6 Ibid.

7 Ibid.

8 Nathan McAlone (14 June 2016), 'Why Netflix Thinks its Personalized Recommendation Engine is Worth $1 Billion Per Year', Business Insider, https://www.businessinsider.com/netflix-recommendation-engine-worth-1-billion-per-year-2016-6?IR=T.

9 Madhumita Murgia (28 March 2016), 'Reed Hastings on Netflix's Global Algorithm', *Financial Review*, https://www.afr.com/technology/reed-hastings-on-netflixs-global-algorithim-20160327-gnrq9m.

10 Mark Butler (15 November 2017), 'How We Made the Demogorgon in *Stranger Things*', inews.co.uk, https://inews.co.uk/about.

11 Lucie Gaget (25 May 2018), 'How Does the Film Industry Use 3D Printing?', Sculpteo, https://www.sculpteo.com/blog/2018/05/25/how-does-the-film-industry-use-3d-printing/.

12 Ibid.

13 Ibid.

14 Angela Watercutter (6 March 2015), 'Drones are about to Change How Directors Make Movies', Wired, https://www.wired.com/2015/03/drone-filmmaking/.

15 Steven Flynn (18 February 2016), 'Drones in Movies: 7 Hollywood Movies Filmed with

Drones', skytango, https://skytango.com/drones-in-movies-7-hollywood-movies-filmed-with-drones/.

16 Ibid.

17 Steve Rose (16 January 2020), '"It's a War between Technology and a Donkey" – How AI is Shaking up Hollywood', *The Guardian*, https://www.theguardian.com/film/2020/jan/16/its-a-war-between-technology-and-a-donkey-how-ai-is-shaking-up-hollywood.

Chapter 23: Work

1 Arthur Goldstuck (18 March 2020), 'How Coronavirus will Speed up the Fourth Industrial Revolution', *The Citizen*, https://citizen.co.za/lifestyle/technology/2257085/how-coronavirus-will-speed-up-the-fourth-industrial-revolution/.

2 John Page (2013), 'For Africa's Youth, Jobs are Job One', Africa Growth Initiative, https://www.brookings.edu/wp-content/uploads/2016/06/Foresight_Page_2013.pdf.

3 NEDLAC (March 2019), 'Futures of Work in South Africa', https://nedlac.org.za/wp-content/uploads/2017/10/Futures-of-Work-in-South-Africa-Final-Report-March-2019.pdf.

4 Joe McKendrick (30 April 2016), 'Does Workplace Automation Destroy Jobs or Create Unexpected Opportunities? An Optimists' View', *Forbes*, https://www.forbes.com/sites/joemckendrick/2016/08/30/does-workplace-automation-destroy-jobs-or-create-unexpected-opportunities-an-optimists-view/#70e953922789.

5 GlobeNewswire (11 February 2020), '67% of Social Media Consumers Aspire to be Paid Social Media Influencers', Yahoo! Finance, https://finance.yahoo.com/news/67-social-media-consumers-aspire-140010515.html.

6 Alex Hern (13 March 2020), 'Covid-19 could Cause Permanent Shift Towards Home Working', *The Guardian*, https://www.theguardian.com/technology/2020/mar/13/covid-19-could-cause-permanent-shift-towards-home-working.

7 Fairwork (2020), 'Second Round of Fairwork's Yearly Platform Ratings in South Africa Launched!', https://fair.work/second-round-of-fairworks-yearly-platform-ratings-in-south-africa-launched/.

8 McKinsey Global Institute (October 2016), 'Independent Work: Choice, Necessity, and the Gig Economy, Executive Summary', McKinsey & Company, https://www.mckinsey.com/~/media/McKinsey/Featured%20Insights/Employment%20and%20Growth/Independent%20work%20Choice%20necessity%20and%20the%20gig%20economy/Independent-Work-Choice-necessity-and-the-gig-economy-Executive-Summary.ashx.

Chapter 24: Rationality

1 Christopher Hitchens (2009), *Letters to a Young Contrarian* (New York: Basic Books), https://www.amazon.com/Letters-Young-Contrarian-Art-Mentoring-ebook/dp/B003P9XC62/ref=sr_1_1?dchild=1&keywords=9780786739073&linkCode=qs&qid=1593501947&s=books&sr=1-1.

2 Reuters (10 October 2018), 'Amazon Scraps Secret AI Recruiting Tool that Showed Bias against Women', IOL, https://www.iol.co.za/business-report/international/amazon-scraps-secret-ai-recruiting-tool-that-showed-bias-against-women-17421247.

Chapter 25: Sports

1 Allen Barra (27 September 2011), 'The Many Problems with "Moneyball"', *The Atlantic*, https://www.theatlantic.com/entertainment/archive/2011/09/the-many-problems-with-moneyball/245769/.

2 Ibid.

3 Michael Lewis (2003), *Moneyball: The Art of Winning an Unfair Game* (New York: WW

Norton & Company), https://www.amazon.com/Moneyball-Art-Winning-Unfair-Game-ebook/dp/B000RH0C8G.
4 Charlotte Edmond (30 April 2019), 'This AI Just Invented a New Sport', World Economic Forum, https://www.weforum.org/agenda/2019/04/artificial-intelligence-invented-sport-speedball/.
5 Ibid.
6 Ibid.

Chapter 26: Memory

1 Kathleen Hennessey (30 July 2012), 'Obama Related to Legendary Virginia Slave, Genealogists Say', *Los Angeles Times*, https://www.latimes.com/archives/la-xpm-2012-jul-30-la-pn-obama-related-to-legendary-virginia-slave-genealogist-says-20120730-story.html.
2 David Nield (14 October 2016), 'Google's AI Can Now Learn from its Own Memory Independently', ScienceAlert, https://www.sciencealert.com/the-deepmind-ai-can-now-learn-how-to-use-its-own-memory.

Acknowledgements

I would like to thank UJ for contributing to the writing of this book. I also would like to thank my former and current students for inspiring me to simplify complex ideas. In particular, I thank Ms Sunita Menon for editing and carefully reviewing this book. This book is dedicated to Nhlonipho Khathutshelo Marwala, Lwazi Thendo Marwala, Mbali Denga Marwala and Jabulile Vuyiswa Manana.